This book is to be returned on or before the last date stamped below

*Swansea Institute
of Higher Education
Library and Learning Resources*

Healthy Buildings

A design primer for a living environment

Bill Holdsworth and Antony Sealey

Longman

Longman Group UK Limited
Longman House, Burnt Mill, Harlow
Essex CM20 2JE, England
and Associated Companies throughout the world.

© Longman Group UK Limited 1992

ISBN 0 582 09388 0

First published 1992

British Library Cataloguing in Publication Data
A CIP record for this book is available from the
British Library

Set by 4 in 11/12pt Compugraphic Baskerville

Printed and bound in Great Britain by
Bookcraft Ltd. Midsomer Norton, Nr Bath

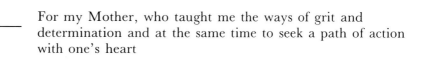

For my Mother, who taught me the ways of grit and determination and at the same time to seek a path of action with one's heart

Contents

State of the Art living (courtesy NOVEM, the
Netherlands)

_____   Do not forget your house.
Here in your own house
You will go about happily.
Always talking together kindly
We shall pass our days.
                    Zuni*

*_Source_ 'Zuni Ritual poetry', 47th Annual Report of the Bursar of American Ethnology (1929–30), Washington DC. US Government Printing Office.

Acknowledgements and author's note

The intention was to describe the current work of healthy building design from all the continents outside the body of Western Europe. In this respect I have not fully succeeded.

Many individuals, institutions, companies and universities were written to; few replied. Some manufacturers and other specialists kept their desk drawers locked. They are not listed.

Regardless of such restraints, this book which is built around my own design method forged in the early 1970s, known as ECHOES (Environmentally Controlled Human Operational Enclosed/External Space), would never have been written had it not been for the patience, encouragement, help and advice offered, and for the abililty to be able to refer to the work of so many who are fully referred to in the text. In particular I wish to thank: Stephen Ashley, Hartwin Busch, Roderick Bunn, John Connaughton, Christopher Day, and The National Union of Journalists in the United Kingdom. Susan Coldicutt and the University of Adelaide, Australia. Bjorn Berge from Norway. Mayer Heilbronn, Germany. Jaroslaw Dowbaj and Wlodzimierz Gryglewicz, Poland. Floyd Stein and Poul Kristensen in Denmark. Renz Pijnenborgh in the Netherlands. Marie Hult from Sweden, and from the United States of America, Dr Sherry Rogers, John S. Reynolds, Dr William Sopper, and the Environmental Protection Agency.

I also wish to thank the editors of the magazines, *Building Services Journal* (UK), *Environment Now* (UK), *Conservation Now* (UK) and the KLM in-flight magazine, *Holland Herald*, and *The Guardian*, where some of the material had previously been published.

A special word of thanks for Ton Alberts' writing of the Foreword, for Antony Sealey's writing of Chapter 3, 'Climate and human life', with such clarity and wit, and for the photographs by Laurence Delderfield. Finally, for Marianne Diebels, who diligently translated technical German and Dutch into English.

Bill Holdsworth

Biographical notes on authors

Ton Alberts: While searching for a cure for a serious illness he began practising Hatha yoga, and then studied esoteric philosophies. It was by this path that he developed a way of defining architecture in a more humanistic way.

Having designed houses, schools and churches for over 40 years, the architectural partnership of Alberts and Van Huut, Amsterdam have now become internationally renowned for the NMB Bank Headquarters, which in 1989 had the lowest energy consumption for any office building in the world. Recent work has included a radical proposal for the building of Europolis — The City of Peace — which followed the earthquake in Armenia. The 'organic' architecture of Ton Alberts will soon be seen in Moscow. In 1990 he formed a new international movement, 'Man and Architecture'.
(Ton Alberts born: 1927)

Bill Holdsworth: He studied building and architecture during the 1939—45 war, but the economic facts of life led to an apprenticeship from fitter's mate to draughtsman in the heating and ventilating industry. Night study and a determination to return to architecture resulted in his working in both the contracting and consulting fields.

In 1967 Bill Holdsworth established his own integrated consulting group whose prime base was building services and energy engineering; it was due to such integration that the design matrix ECHOES was developed.

His professional colleagues often suggested that his ideas were before their time. Widely known as a writer, journalist, lecturer and broadcaster, he at present lives and works in The Netherlands.
(Bill Holdsworth born: 1929)

Antony F. Sealey: A native of Worcestershire, after service in the RAF Antony Sealey qualified in architecture at the Birmingham School of Architecture. After working in official and private offices, he took up full-time teaching at the Birmingham School specializing in science, technology and history, with particular interest in climate and materials. Elected Fellow of the Royal Meteorological Society in 1962, and Honorary Senior Fellow of Birmingham Polytechnic in 1985. Now teaches part time, and enjoys his garden.

Publications: *Bridges and Aqueducts*, Evelyn, 1976; *Introduction to Building Climatology*, CAA, 1979.
(Antony F. Sealey born: 1927)

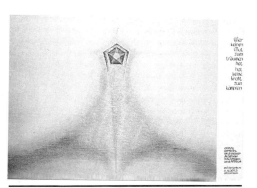

Ton Alberts' motif: Those without the courage for dreaming have no strength to fight . . .

Foreword: buildings are for people

I am honoured by the suggestion that I should provide a foreword for a book on a subject which is near my heart. As an architect I participate in a world which asks for new houses and offices or is asked to restore old ones. To design a building is answering a question which goes beyond the mere necessity to provide for people a roof over their heads.

We live in buildings and work in them. We feel happy or sad in the building in which we remain. A building is as a third skin. We have our own skin, most of the time we put some clothes on a second skin and our built environment can be seen as our third skin. In other words, a building has to be as comfortable as our skin; it lives, it breathes, it surrounds us without oppressing us. A building has to be as healthy as we want our own body to be.

From this point of view it is easy to understand how strongly we can be influenced by buildings. A sick building makes us literally ill, where a healthy building makes us feel better. A healthy building shows its beauty to the world. This beauty is more than a successful aesthetic design. We touch buildings every day, feel the character of used materials in the doorknobs, the banisters. It invites us or rejects us. A healthy building feels better, communicates with the world and the people who live and work in it as if it were inviting people to enjoy its warmth and conveniences.

Sir Winston Churchill once said: 'We shape our buildings; afterwards they shape us'.

High-tech buildings emphasize the rational part of us. We become over-rationalized. Although new construction techniques and advanced instruments are used, it seems that we have forgotten that we also have to live in such buildings. Besides this rationality we also need fantasy, intuition, and emotions.

Yet, I trust that we are envisaging a world in which the time is changing! When we are aware of the reciprocal relationship between the buildings we create and the face of this world, we must then build for this world with renewed perspectives and insights.

Bill Holdsworth understands this very well. As no other writer on architecture, he is familiar with this subject. The book now before us provides a complete overview of the developments in healthy building design and offers us a thorough investigation of the new ideas and principles in the design world. Most helpful, too, will be the illustrations, which are drawn from a wide area and illustrate the subject in an intelligent way. The advanced

Ton Alberts (photograph by Bill Holdsworth)

study of several projects which are carefully compiled and up to date makes this book a valuable design primer.

I can recommend this book to all architects, students and those of the general public who are seriously interested in healthy buildings. May it enjoy a broad range of readers with whom we in joint effort can create a vital world. A world that leads us into beauty.

TON ALBERTS

1 Introduction: early thoughts and actions

It was some forty years ago that I began to realize that to engineer a comfortable environment within a built space it was necessary to understand the external forces created by both nature and man.

War-time education at London's Northern Polytechnic School of Building not only taught me the four powerful crafts of bricklaying, carpentry, plumbing and decoration, but had also introduced me to the elements of pollution in a subject called 'sanitation'. Here we learnt that a hospital should not be placed next to a graveyard, or a housing estate close to a chemical plant. But such conjunctions were rife, and I wished that the bombs showering down onto our cities would remove the dross of pollution so that my student dreams of seeing the open land of places like Hampstead Heath extended to reach through the slums of London might be fulfilled: great lungs of fresh air from the countryside beyond the city to oxidize the banks of the River Thames. During my apprenticeship in the heating and ventilating industry, Mr Black the chief draughtsman at J. Jeffreys and Company, had ensured that both heat loss and solar gain calculations were painstakingly made for all aspects of a building.

I had wanted to be an architect, but the social and class system of pre-war England had conspired to prevent such an idea, along with the alternative of being in films. Nonetheless, it was whilst I was an apprenticed engineering draughtsman that I began my life-long interest in architecture, the arts, and the way that people lived with their own unique interpretation of being at one with their surroundings. In becoming a building services engineer, that person who has to knit together all the different physical sciences in a way that results in the creation of a comfortable internal environment, it was clear to me that one had to be both an artist as well as an engineer. To express such ideas was grudgingly accepted. But to put such ideas into practice was strongly opposed by engineers, architects and builders alike.

Although much is now being talked of healthy building design procedures, there are still some architects who consider that the building is purely their preserve. In what is becoming known as 'biological architecture' people are also emerging, particularly in Germany and the Netherlands, who consider themselves as priests of this new movement, interlacing concepts of eastern philosophies and ways of perception. I see nothing wrong with such ideas provided that they are related to the practicality of living in a mechanized and electronic world, able to find a way

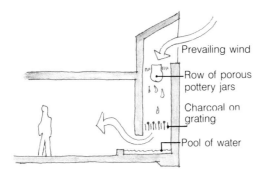

Fig. 1.1   Evaporative cooling for rest centre, Tel el Kebir, Egypt. (Invented by the author, 1951)

to ensuring that we all understand how to manipulate the forces around us.

The sayings from the Navajo and Pueblo Indians of the south-western United States that introduce each section illustrate our being part of nature in a simple and direct manner.

The first time I tried to find a way to use natural means that would not damage the immediate surroundings was as a National Service soldier serving in the Egyptian desert a few years before the Suez Crisis. At Tel El Kebir, our fresh water came from deep beneath the desert sands, pumped to the service from deep bore-holes. Water for washing and cooking was heated by burning derv engine oil, which produced a thick black cloud which suffocated all the people working in the cookhouse; globs of burnt oil would be mixed with the water, making the tea awful to drink and washing a messy business. Necessity as the mother of invention may be an old saying, but it never loses its truth. By cannibalizing a finned tank engine cooler a rudimentary solar water heater was built. It proved its effectiveness when the clean hot water it produced nearly took the skin off the Sergeant's back.

Living in a tent in a hot dry climate had many advantages, but to find a cool place to rest, to study and 'get away from it all' was a popular demand. An old, partly demolished, mud building was found. Using techniques learnt from a study of old Arab buildings in the area, the walls and roof were built to keep out the intense sun yet at the same time allowing daylight and cool breezes to enter the interior. An inner yard was formed with a canvas shade. To cool the inner room a row of porous pottery jars known as 'chatties' were set high upon a platform. In these, water would slowly evaporate. From a corner tower, hot prevailing winds were encouraged by a canvas mill-sail to enter the tower shaft. The hot air would flow over the cooling jars, a charcoal filter and a pool of water at the tower base helped to reduce grit, and the internal space was thus cooled by natural air conditioning.

I make no apology for this design primer for healthy buildings having been written and compiled by a building services engineer. It does not matter who introduces the new primers for global perception provided that we don't look upon such ideas as another 'flavour of the month'.

Writing in *The Guardian* in June 1966, I talked then of a 'Blueprint for a Better Life'. Speaking of the engineers who could achieve such a blueprint, I wrote, 'Now the designer of mechanical and electrical services requires a knowledge of several sciences together with an appreciation of architecture, building methods, structural engineering, and a change of hats to see him through! ... In time there will be men and women able to combine and control all these talents'. These Environmental Engineers as I then called them would lead action planning teams to '... deal with problems of fuel, distribution of power and ancillary services, together with techniques of building with components capable of responding to rapid changes in population and communication ... buildings must be muffled against noise ... that we should find ways to trap the sun's energy for domestic and commercial use ... houses could become vastly different in shape, determined by all the environmental considerations ... We have the ability and scientific knowledge to control our environment yet often lack the will to think imaginatively ... The climate of our lives can be transformed if we divorce traffic from housing, allow for filtering unwanted noise, and build to resist the rigours of our changeable weather'.

I concluded by saying that 'Such a programme should not

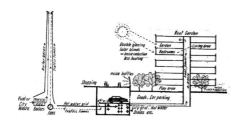

Fig. 1.2   Illustration for 'Blueprint for a Better Life' by Bill Holdsworth

be beyond our reach in these days of nuclear power and flights into space; it would give city sense to city planning ranging from a room to a region'. The words written twenty-five years ago are as pertinent now as they were then, except that the factors of pollution that now influence the way we live on this planet have increased with time.

Antony Sealey, the well known climatologist, has joined me in asking that not only architects, engineers and builders make health a prime design element, but that all decision makers should become aware of these issues.

The clients who prepare a brief for a new town or a new building should make it their responsibility to ensure a healthy living environment. The same responsibility rests with our representatives of local government, as well as every thinking person.

I am conscious that in Chapter 4 I have paid scant attention to the importance of colour, the addition of art objects and sculpture within the built space, and the interlinking of architectural space and objects of human use and comfort that results from the designing of the interior space. These elements are also important in contributing to our sense of inner health, aids to our ability to cure ourselves. It is not that I feel them beyond my ability to present them; it was the knowledge that readers themselves should also be able to take part in the creation of a new architectural form of thinking. An organic form where we need to move through the colour and other perceptive artistic senses as to the way we build both our cities and our buildings. In the colour spectrum we find that architecture is often related to the active colour red, moving through the structural to social life. From the hard to the soft. From red through purple, blues, yellows to the green of poetry and dance. Senses that might bring a smile to many engineers. When my office colleagues learnt in the mid-1960s that I was engaged in modern dance ballet, they produced a poster with the slogan that I was the only engineer with a pair of ballet tights in his brief case. Later I found that my dance experience stood me in good stead as a consulting engineer working on theatre projects. I understood the life of the dancer and would ask questions that resulted in providing an environment sympathetic to the world of the artist. Good design is also good business.

Ton Alberts, the radical Dutch architect of organic design thinking, states quite simply that buildings are for people. We have now, in the 1990s, started to call such concerns a 'greening'. From some early thoughts and actions this book is a contribution to ensure both a greening of architecture and the engineering services that makes for a living environment.

2 Health as a design element

My words are tied in one
With the great mountains,
With the great rocks,
With the great trees,
In one with my body
And my heart.

Do you all help me
With supernatural power,
And you, Day!
And you, Night!
All of you see me,
One with this world!

                              Tewa*

The Alhambra Palace rests easily on the high rocky escarpment above the dusty city of Granada. From its red sandstone walls the snow-capped Sierra Nevada mountains silhouetted against the eastern morning light come close to an embrace. A natural balance has been created by man with nature.

Within the walls of this great Moorish solar palace, sheltered gardens and inner courtyards, complemented by graceful fountains, cool and temper the hot summer winds; havens of calm, subtly designed, allowing the inhabitants to catch the warmth of winter sun before entering the deep inner chambers where steam baths smooth away the ache of cold bones.

To reach the palace you can walk slowly uphill from the Plaza Nueva through the narrow canyon of the Street of Guitars. The last climb is through an avenue of sweet-smelling trees. To enjoy this journey it is advisable to wake early, because shortly after the tourists have had their breakfast, giant coaches noisily grind and crawl up the hill, filling the street of guitar-makers with suffocating toxic exhaust fumes. The air is poisoned both for humans and plants. The trees are barely able to re-oxidate the air. Door-to-door tourism can be decidedly unhealthy!

But if you pass through the guitar-makers' workshops you enter a space of peace and quiet: cool inner courtyards with their mixture of water and plants temper the climate of nature. The spirit of those medieval Moors now also has to cope with a manufactured poisonous climate far more volatile than anything nature seemed to present.

One answer would be to ban the coaches and to build at another spot a funicular railway. Whether it would end the doubtful prosperity of the guitar-makers is open to question. All solutions have a price.

Granada, Spain, 37 10 North 03 35 West
12 June 1990

Living in a sick building

The preparational work for this design primer was carried out on the ninth floor of an apartment block, constructed in 1980 of high-density concrete open-sided cubes, with elemental end-wall thermal insulation, a flat roof and large glass windows (single glazed in the bedrooms) set into aluminium frames.

There was space enough. Large open balconies afforded extensive views of the university city of Nijmegen close to the Dutch−German border; yet to open a window immediately brought into the living space the excessive noise levels emitted from heavily laden trucks speeding towards the motorway link with Europe, and, worst of all, the high-pitched whine of small motor bikes. Even without the noise, keeping the windows open

*Source Herbert J Spinden *Songs of the Tewa*. New York Exposition of Indian Tribal Arts, 1933

on days when high winds sucked and blew against the large areas of glass caused havoc to normal comfortable living.

Closing the windows effectively sealed the room space, there being no mechanical ventilation system. Radiators with marginal control were difficult to adjust to suit body needs. Temperature gradients produced cold feet and warm heads. On a hard concrete floor I had placed an affordable synthetic carpet covering which gave off sparks of static electricity and a strange smell. On entering the apartment after being away for some days I would immediately start a bout of uncontrollable sneezing.

Long before I extended my research into chemical and electromagnetic contaminants, I had already begun to realize that I was living in a decidedly unhealthy building. Not only was the building constructed of energy-wasteful materials and in itself a heat sink, but also the components of construction and furnishing were detrimental to my health.

Along with my neighbours I was living under a vast Faraday Cage. I purchased an ionizer and placed it in the room where my PCW was stationed. A white sheet of paper placed under the ionizer soon illustrated the high level of dust and other airborne particles that inhabited my immediate surroundings. To combat the hard sterile surfaces I introduced many varieties of plants.

Environmental sensitivity is usually manifested in the major organ systems, especially those related to the smooth muscle (respiratory, gastro-intestinal, neuro-cardiovascular, genito-urinary) systems and skin. There are usually a variety of symptoms, depending on the specific toxicity of the substances, the severity of the exposure, the number of organs involved, and the patient's individual susceptibility.[1]

Diagnosed case studies of such chemical hypersensitivity are fully illustrated in Section 4.2.

There is a period of opportunity, between optimum health and indentifiable end-organ diseases, when one can reverse or prevent end-organ failure. This time span can occur days, weeks or years before a recognized disease occurs.[1]

It has been estimated that through our changing habits the majority of the world's population spends some 85–90% of its time living in an internal environment of one form or another, the greater part being buildings, whether factories, offices or homes. With the rapid increase in world population, especially in Asia, Africa and South America, the majority of people will find themselves confined to vast cities that are speedily erected with little dignity and only a passing nod to planning for health. One such example is the growth of Lagos in Nigeria. A country that once supported a well-developed agricultural self-sustaining society has now, through a shift to an industrialized city culture based on the exploitation of oil, become a nation totally dependent on food imports. The pressure of high-density population in cities such as Lagos has led to a rapid decrease in the nation's health.

A typical North American city of 100,000 people imports 200 tons of food a day, 1000 tons of fuel and 62,000 tons of water per day. On the output side, these same cities dump 100,000 tons of garbage and 40,000 tons of human waste per year. Mobility and fragmentation with the acceptance of long distribution systems is the price we pay for what is often lovingly referred to as 'progress and development'.

Until June 1990 the city of Philadelphia sent its incinerated ash across State lines for disposal in Ohio, where it is now transported 'cross-border' to Panama.[2] Similar world-wide transportation of hazardous materials is being increased. Networks can dump nerve gas from Germany to be incinerated in a Pacific atoll, while others trundle spent plutonium rods from Japan. These are shipped to the port of Dover and then taken by overnight train across heavily populated cities such as London and Birmingham, where another reprocessing operation will exhaust lethal contaminants, albeit in small quantities, that will build up in our external atmosphere and be spread by prevailing winds across England and the continent of Europe.

In the past 15 years the cost of disposal of one ton of garbage has risen ten times, whereas in the same time the cost of disposal of one barrel of hazardous waste has risen by 300%. Contractors still spill the contents on the side of a road; and as the laws for environmental control tighten in those countries where the will of the people is more respected, the danger can be that the waste of the rich nations is foisted on to poorer nations.

So that we can build healthy buildings, we also need healthy surroundings. Although this book includes examples and pointers for ecologically sound approaches to the planning of urban space, as with many other subjects that will be referred to, space limits a fully comprehensive survey.

In 1984, the World Health Organization concluded that between 10 and 30% of all buildings constructed throughout the world were sick, and that the exact number of people affected by these buildings was unknown.

The number of different chemicals and synthetic substances used for building materials was about 5000 in 1987, but such materials are increasing[3] and at the same time the different pollutants found within the built space emitting chemical, radioactive and biological compounds are mixing into a cocktail that is becoming increasingly difficult to control and analyse.

The cocktail effect

The sick building syndrome (SBS) is not new. Long before the advent of fast-track construction, sealed interiors, and the use of synthetic materials, many buildings constructed of traditional materials were also unhealthy. Older illnesses derived from dampness and condensation, and still persist. In many industrial cities the prefabricated concrete panelled tower blocks built to replace the Victorian slums have become slums themselves, made worse by being built vertically into the sky. A study in 1972 of tower blocks in Paisley, Scotland, illustrated the inability of the architect or building contractor to design for climate as well as for health. The timber frames were of softwood with single glass panels. The external wall was pebble-dash, little chips of gravel set into a cement render that covered concrete blockwork. It was a method beloved by speculative builders of the 1930s when tudor-style semi-detached houses were built alongside the new arterial roads in Britain. The same construction technique was used for tower blocks, that took the full wind force of Atlantic winds, which held sea-salted rain close up to the external walls. The rain cut through the walls like a knife through butter. The result was condensation and dampness. As a consequence of inadequate heating and ventilation both the buildings and the people living in them became sick. Vast sums of money were spent by local councils to encapsulate the buildings (see also p. 34). The best solution would have been either to demolish them, or to spend even higher costs in a programme of urban retrofitting. Housing, like health, seems to become the first victim in national budgets when industrial and business economies decline.

The Paisley investigation took place nearly 20 years ago. But little has changed. Similar illustrations can be cited in Moscow, Warsaw, the USA and other countries where climate and health were disregarded during the building process.

As with the introduction of the Public Health Acts in England during the years of the first Industrial Revolution, it was only when plague and cholera manifested themselves upon the more affluent members of city societies, that legislation for environmental health was instrumented.

The recent hotter summers, coupled with a breakdown in garbage collection, main-line sewers needing repair, and the disruption of rats form the old London Dock areas, has resulted in an increase in the rat population which swarms the streets not only of the poorer boroughs of London, but also finds that the well-conditioned interiors of high-cost accommodation offer an equally comfortable abode.

International companies such as Rentokil (see Chapter 8) not only offer services to improve internal living environments with tropical plants or ways to combat Legionnaires' Disease by eliminating the baffle jelly that accumulates in plumbing systems (see Section 4.2), but they also use hi-tech methods to get rid of rats and other pests.

Until quite recently employers and staff management in the USA, Japan and Western European countries dismissed SBS as an excuse for not wanting to work. Slowly we are beginning to learn that this was not so. For the human body, as well as the world's eco-system, there appears to be a weakening of the immune defence systems.

Insidious and difficult to account for are the viral infections that affect 30% of people in Western Europe, making it one of the greatest sources of illnesss. Man's activities, including his industrial, transportation, and energy systems, affect not only the global climate but also the quality of the air he breathes within his places of habitation.

Carbon dioxide has, over a number of centuries, been trapped in ice-bubbbles. Measurement of such bubbles has given research scientists an indication of the comparative fluctuations of carbon dioxide in the atmosphere. It would seem that over a period of 160,000 years fluctuations stayed within a range of 200−300 ppmv (parts per million by volume). The highs and the lows were affected by changes in the earth's temperature.[4]

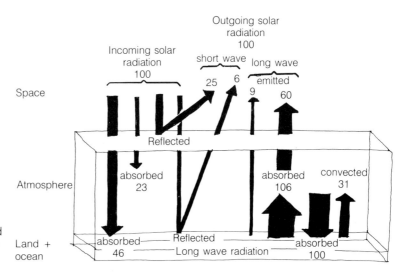

Fig. 2.1 The Earth's energy budget, scaled to 100 incoming and 100 outgoing arbitrary units of solar radiation

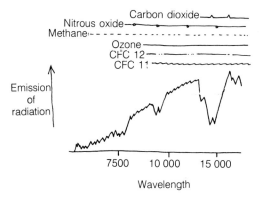

Fig. 2.2   The energy emitted by the earth in different wavelengths, showing the wavelengths absorbed by the principal greenhouse gases

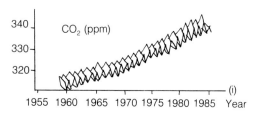

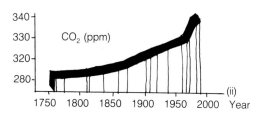

Fig. 2.3   Increases in atmospheric CO_2 levels recorded in the air at (i) Mauna Loa and (ii) in ice bubbles. (After Cannell, M.G.R. and Hooper, M.D. (1990) *The Greenhouse Effect and Terrestrial Ecosystems of the UK*, Institute of Terrestrial Ecology, No. 4, HMSO.)

Ice core data showed that, in the 1750s, the concentration in the atmosphere was 275 ± 10 ppmv. But from the time of the first Industrial Revolution there has been an indisputable increase from 315 ppmv in 1958 to 350 ppmv in 1988. Carbon dioxide concentrations are still rising, and are now higher than at any other time over the past 160,000 years.

The authors of the Institute of Terrestrial Ecology research publication No. 4[4] inform us that in presenting a simple picture for our initial understanding, there is hidden an enormous complexity with some considerable uncertainty about the magnitude of the carbon fluxes.

Carbon dioxide is one of the main gases present in the atmosphere contributing to 'the greenhouse effect'; the other gases with similar properties that tend to be transparent to incoming short-wave solar radiation, but absorb outgoing infra-red radiation, are methane, nitrous oxide, ozone and certain chlorofluorocarbons (CFCs). The resultant changes in climate from the greenhouse effect will and are having a bearing on the way in which we must build and condition our internal environments.

We build habitations to protect us from the ravages of an inclement and volatile natural climate. But we are not only raising the temperature, we are also adding other man-made pollutants to what we call, so beautifully, 'fresh air'.

Into this 'fresh air' we add daily ever-increasing amounts of volatile compounds. The exhaust from cars and lorries, sulphurous discharges from factory chimneys, radioactive emissions that seep into earth, air and water that will persist to cause dreadful harm to the human body some thousands of years hence. Each emission adds to the greater sum. Other electromagnetic forces disrupt the original natural balance. These are the external factors of influence (Section 4.1) swirling around the outside of our buildings.

Internally we are also allowing ourselves to suffer unnecessarily. Many volatile organic compounds are given off by building materials, furnishings, consumer products, office equipment and supplies. Many of these compounds have known or potential health and physiological effects. To these compounds are added other chemical pollutants, as well as non-chemical factors such as electromagnetic radiation, biological products, dust mites, and social actions such as cigarette smoking.

It could well be said that our living and working environments are becoming a carcinogenic soup. The sick building syndrome is not just a matter of inadequate ventilation rates within tightly sealed buildings, supposedly protecting us from the extreme of our external climate, but a whole cocktail effect of natural and

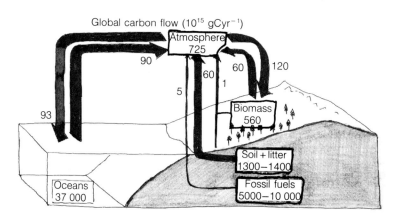

Fig. 2.4   A simple global carbon budget. (Note: the numbers are approximate.)

man-made contaminants. Clearly, man-made contaminants are tipping the natural balance.

Given all the negative external forces and internal components, the task before us is to find a way to ensure that we can build a reasonably healthy building. I use the word 'reasonably', because in the overall equation of the built construction there must always remain a realistic assessment of materials that are both available and affordable, and synthetic materials should not be rejected simply because many have proved to be either unhealthy in use or are wasteful in terms of global energy use. Each material must be judged on its merits. The first question should be to check that its use will not harm us.

A building is like a person in that its components grow old and decay. In the same way that we stay healthy by keeping fit and clean, so also a building must be kept clean, flushed out on a daily basis, and subject to periodic checks for repair and maintenance. Like ourselves, a building needs to be cosseted. Without proper care a healthy building will soon become sick.

ECHOES: a design matrix for health

Wondering why thick moss grew on the roofs of some old people's bungalows in the English steel town of Corby led the author to the development of an environmental design matrix that took you from a study of all the external climatic and natural surroundings into the internal space to be conditioned.

The year was 1970, and the first Energy Crisis was yet to occur. but as a building services engineer engaged in a design for a district heating scheme I wanted to find ways to limit the amount of fossil energy used, and to establish a heat-rent affordable to the occupants of the local authority housing that was to be built.

An elderly lady who lived in the bungalows told me that water lay a few inches beneath the surface in the winter where the air temperatures dropped to $-10°C$. Such local information came as a direct contradiction to the informed opinion of the day. Both Parker Morris Standards issued by the Government and the codes of practice of the Institution of Heating and Ventilating Engineers (now the Chartered Institution of Building Services Engineers (CIBSE)) recommended a design base outside winter temperature of $-1°C$.

Taking local knowledge to heart led to a study of the immediate micro-climate.[5] This resulted in a new awareness that external winter temperatures could even reach $-15°C$, that the wind did strange things due to the creation of landfills and man-made hills, and that the very landfills were poisonous and affected the ground-water conditions.

The primary aim was to achieve an energy-conscious building system, with cutbacks on the use of fossil fuel and also to introduce simple concepts of renewable energy technology. The latter were ignored. The client-body, in this case a local council, were yet to be convinced of such untried methods.

The scheme was successful in respect of the saving of energy. The houses were healthy, being constructed in traditional materials, and the heating system allowed for openable windows.

Slowly this system of design grew to include many other areas of concern, at that time deemed to be outside the remit of most people who formed a design team, whether architect or engineer.

During my war-time (1942–45) training at the Northern Polytechnic School of Building in London one of the class subjects was sanitation. This subject dealt not only with drains, but we also learnt to be aware of the contaminating effects of graveyards, and the positioning of hospitals and factories, to check the direction of wind, and to be aware of cross-contamination from places that were areas of sickness. These sick areas also included toxic wastes form industrial plants. The new plan for London produced in 1945 clearly showed how cities should be divided up between areas of living space and areas of industry.

A wonderful plan for my area of north London showed Hampstead Heath being linked to Primrose Hill, Regent's Park and on to Hyde Park: huge lungs of countryside being led into the city. What wonderful dreams we had then!

It was those lectures in sanitation that stayed with me throughout my working life and I saw that they were the basis of healthy city design that had stemmed from the Public Health Acts of the late Victorian city engineers.

During the design period for the Charles Wilson Building at Leicester University, in 1967, the architect Sir Denys Lasdun told me that he wanted nature to come from the outside, to move through and touch all elements of the building space, and then to move back to the outside again, without being fouled up by mechanical services. It was this concept of moving from the outside in, then back to the outside, forming a complete circle, a wholeness, that eventually led to the development of ECHOES (Environmentally controlled human operational external/enclosed space).

This book makes use of the developed system ECHOES (environmentally controlled human operational external/enclosed space) and examples are given of its early use. Taking a roll of tracing paper and starting from the outside of a proposed building complex, ask probing questions on the natural and man-made climate: what are the facts regarding pollutants including noise, fumes and toxic ground conditions, the positioning of overhead electric cables, and the type and use of adjoining property. Then enter on the architectural sketch plans of the enclosed space, determining functions of use; ask what materials are to be used and furthermore question even some of the assumptions made by the architect or other members of the design team.

Such a design method allows for the best building form to be determined both visually and also in terms of climate, energy, shape and people's use. It is in effect the undertaking of an environmental audit where alternatives should be proposed to clients to counter any environmental health risks that may emerge.

There are many technical solutions, which in the right context, correctly executed and employed, will give a satisfactory result. Some solutions are more sensitive to the commonest faults found in planning, construction and occupational use. Cutting costs at the design stage in areas that would benefit from more intensive investigation only leads to risk solutions.

When risk solutions are employed, inspection at different stages of design, installation and commissioning becomes even more exhaustive and tends to be left out in the desire to meet the conceived building completion date. Such actions lead to trouble and increased costs. It is better to avoid risk solutions in the first place. The final tuning up of any building is as important as getting yourself prepared for a sports event in which you are participating.

Clients will have to realize that design fees must be expanded to include more people in the design process, such as occupational health workers, doctors, specialists in toxiciology, as well as the men and women who are to live and work in the building. The final result is cheaper in the overall cost equation both structurally and from the human point of view.

Within the architectural professions one hears special words bandied about, sometimes whispered in a priestly way. Discipline: a maintenance of order, a system of training for orderliness and efficiency. A new favourite word is parameters: a constant factor or variable which determines the specific form of a function. Such words are important to give us some order of approach. ECHOES is in itself the bringing together of disciplines and the application of parameters. But I offer a word of caution. There is a tendency in a number of European countries to compartmentalize the study of building physics, to keep apart the different faculties of architecture, engineering and inter-active subjects. I have been painfully aware of this in the Netherlands, where imaginative legislation in urban planning, climatic awareness and building physics, which produces the most 'organic' way of bringing together the many variables that need to be looked at in the design for healthy buildings, is being eroded by the methods of teaching.

There is a need for a harmony of understanding of the parameters both in the classrooms of Schools of Architecture and in the Schools of Building and Engineering. This same organic approach must follow on into the design office, specifications, codes of practice, the shop floor, construction sites and inspectorate both as part of the built operation and the constant monitoring by government and other institutions of the more immediate and expanded global atmosphere.

As yet there are no ready-made answers on how to dispel completely the cocktail effect except by limiting the contaminants in the first place and by legislating for an ecologically sound approach to urban space. It will require commitment and the political will to spend our resources in a healthy and holistic way.

As the designer or decision-maker, you have a responsibility not only to yourself but also to all the people who live both in your garden of activity and in adjoining gardens of living space.

The information contained in this book is not usually found in the normal textbooks that form a designer's toolkit. The list of materials and other evaluations is just a beginning. It will be your task to constantly update the information.

The intention for this book is to be seen as an illuminator, a pathfinder towards ensuring that we build with a sense of fully understanding that we live on a fragile earth and in a climate that never stands still.

Nijmegen, The Netherlands. 51 50 North 05 52 East

3 Climate and human life

Antony F. Sealey

The Earth is looking at me;
she is looking up at me
I am looking down on her
I am happy, she is looking at me
I am happy, I am looking at her.

The Sun is looking at me;
he is looking down on me
I am looking up at him
I am happy, he is looking at me
I am happy, I am looking at him.

Navajo*

Definitions and records

The intention of Chaper 3 is to clarify some important concepts in the science of climatology, particularly as they affect human survival and relationships with elements of the human habitat, which may be both physical and psychological. A distinction should first be made between climate and weather: although intimately related, they are not identical. A good analogy explaining this distinction was made by a well-known climatologist, C.E.P. Brooks,[1] in the mid-1950s. As defined correctly, weather is the total condition (often separated for convenience into elements) of the lower atmosphere as it can be experienced or measured with appropriate instruments at an instant. Climate is the summation of weather conditions at a particular place totalled or averaged over many years. The analogy suggested by Brooks considered a similar distinction between money in your pocket at an instant and your bank balance at the end of the year; they are not identical but are closely related. Fluctuations in weather, given that there will inevitably be good and bad days, are smoothed out by averaging over a period.

There are, however, some problems. For while a designer hoping to assess the climate of a potential site for, say, a new town may well find existing records inadequate, these cannot be expanded at short notice; some inferences may be drawn from weather conditions about the likely climate of an area — rather as we may be tempted to make a judgement about a place not familiar to us on the strength of a one-day visit. This would be scientifically unsound since it might have been a particularly good day, or, conversely, a bad one. A related problem which this raises is that because climate, as well as weather, tends to change with time, a collection of data made over a period of 30 years becomes out of date while it is being collected, but at least, together with weather records and local knowledge, it can give information about trends which preceded change and those which did not.

The biosphere

The human animal requires supplies of oxygen, water and energy foods in order to survive; furthermore, whenever or wherever the species emerged it must have done so into an environment that favoured its survival with the minimum of technology. The components of that environment — or any other in which living

*Source Navajo Creation Myth. The story of emergence. Santa Fe, New Mexico Museum of Navajo Ceremonial Art, 1942

is contemplated — must be understood before we consider modifying it, or, for that matter rescuing it, since it has probably already been modified considerably.

A general term for the whole of the natural, physical environment in which life occurs is the biosphere. Although life as we know it can be artificially sustained outside this regime it is normally found within a zone surrounding the earth and consisting of the oceans, the upper crust of land masses and part of the atmosphere — itself an envelope of mixed gases which is divisible into a number of layers. In the lowest of these, about 10 km thick, the troposphere, the phenomena collectively known as weather events occur, although they may well be influenced by the nature and modifications of the upper layers. Typically the composition of the atmosphere may be described as follows:

	Percentage by volume
Nitrogen	78
Oxygen	21
Argon	0.9
Carbon dioxide	0.03 < >
Neon/helium/methane ⎫	
Krypton/hydrogen/ ⎬	less than 0.10
Nitrous oxide/xenon ⎭	
Ozone	variable
Water vapour	variable

plus traces of other gases such as radon (emitted from the earth's crust)

The total thickness of the biosphere may be regarded as about 15 km, from about 5 km below sea level to about 10 km above, although this varies with time and latitude.

If visualized on a scale model it gives a good idea of the tenuous nature of our habitat, which should worry us! On a large schoolroom globe, say 800 mm in diameter, the total thickness of the biosphere would be represented by 1 mm.

Elements of weather

The elements of weather, such as rainfall, air temperature (average, maximum/minimum, highest/lowest) are susceptible to measurement when they become the stuff of which records are made, whether by hours, months or years. A growing selection of instruments is available to enable two different kinds of data to be collected, and two correspondingly different families of instruments can be recognized by the endings of their names.

An instrument ending in '-ometer' normally indicates that we are concerned with an instantaneous value, while an '-ograph' ending suggests a meter which records over a period. Both kinds of instrument have their uses, as, indeed, have both average and extreme values, whether maximum or minimum; it all depends on what we want to know.

Scales of climate

Rather in the way that time scales for weather measurement may vary, so may we be interested in the climate of different-sized areas. The oldest map of world climate, Sacrobosco's *Sphaera Mundi* of the late Middle Ages, indicates very simply habitable as against non-habitable regions, which we would call climatic zones (polar, equatorial, and temperate). These simple divisions

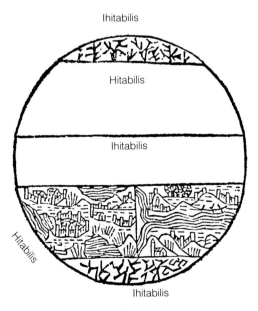

Fig. 3.1 'Sphaera Mundi'. A very early description of the Earth's climatic zones, 1230.

are frequently too crude, particularly since we are likely to distinguish within each major zone a wide variety of local climates, often due to topographical features such as hill ranges and lakes. Not only may the hill range have a different climate from the plain out of which it rises, but also if it runs north—south it will exhibit very different local climates on its two sides.

Then if we examine local climates in finer detail, we are likely to discover very small-scale variations (e.g. the classic distribution of moss growth around the trunk of quite a small tree). These scales are generally referred to as:

1 Macro: climatic zones and major areas (continents)
2 Meso: smaller regions or districts
3 Micro: the smallest scale: a loose definition which would depend on what the observer's interests might be

Some decades ago, the term curtain walling came to mean a thin, impermeable space divider system. There were many attempts to create this thinnest of boundaries which created as many problems as opportunities. Sometimes the solution was seen to be a single glass sheet in lightweight metal mullions, cleverly extruded to provide fixing for the glass with spring-in metal beads and gaskets. The theory was simple and derived from patent roof glazing, which had been around for about 100 years. The glass could be clear for lighting and vision, and perhaps colour-enamelled for privacy, with wire mesh rolled into the glass of safety.

Construction was simple and rapid. There was, however, a micro-climatic defect that was not predicted. The mechanism of this seems to have been that unless the incident sunlight was reaching the glass squarely, normal to its surface, the mullions would cast shadows on one side, i.e. along one edge of the glass sheet, giving it in effect a significantly different climate. Solar heat would build up in the opaque glass, causing cracking parallel to the mullions due to differential thermal movement. Furthermore, the wired glass turned out to be weaker than unwired glass since the mesh provided a shear plane halfway through the thickness. Investigation of freezing cold internal temperatures at Stirling and Gowan's Engineering Building at Leicester University in the winter of 1960 found that the culprit was the glass curtain walling. As the wind pushed and sucked the glass within its metal frame, snow and cold air were allowed ingress into the building. Micro-climates can exist at even smaller scales, and can be the cause of annoying rather than dangerous effects, such as pattern staining.

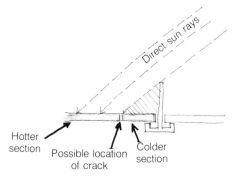

Fig. 3.2 A micro-climatic effect: a patent-glazing bar will cast a shadow on the cladding material — colder section to hotter section differential Δt°C.

Hotter section   Possible location of crack   Colder section

Classification of climates

At the larger and middle scales there have been many attempts to classify climates, often in terms of temperature and moisture records, for example the systems due to Köppen and Thornthwaite.[2] While some of the systems are complex, they help to make the point that the elements of weather and climate do not occur singly, in isolation, although they are often studied, measured and recorded in that way. Although many climatologists must have been aware of this interaction, specialists tended to concentrate on a single element. In recent years in particular the composite nature of weather has come to be accepted as important. R.E. Lacy in England realized the significance of the observed fact that heavy rain and high winds often happen together. He studied directional rainfall and

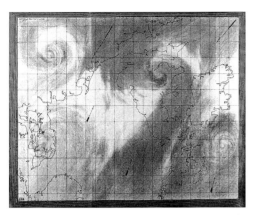

Fig. 3.3 Typical collision of warm, wet and cold, dry airstreams over the North Atlantic (photograph by A. Sealey).

eventually derived an Index of Driving Rain (BRE, 1971).[3] This is an important concept if we are interested in the permeable nature of exterior walls.

Interaction of elements

A somewhat similar coupling of elements forms the basis of an idea developed from studies of human endurance (the freezing of US troops out of doors) by Siple and Passel in the 1940s.[4] The term wind-chill which they coined has recently been accepted into the folklore of the weather forecast. It describes a well-known phenomenon, that a simple phrase such as, 'It feels colder today because of the freshening wind', covers the multiple effect on the human body of a number of elements, probably including air temperature, air movement, humidity and the presence or absence of incoming radiant energy. 'Feeling colder than yesterday' may mean that the temperature is lower today, or that the temperature is the same but heat is being removed more quickly because of air movement, or yesterday was sunny but today will be overcast, or variations in all these factors acting together (see Chapter 5, Case study 3).

There have been attempts to measure all these elements with a single instrument: a very complicated operation. However, one suggestion is that a normal mercury thermometer standing in a black-painted metal can filled with water (approximating to the human body!) may give a very useful model of the multiple components of thermal comfort.

Sources of power

The restless atmosphere is a somewhat poetic statement of an important fact. The atmosphere is neither stationary nor uniform or unchanging in its composition. Indeed, meteorologists who, like most other physicists, would prefer to be able to rely on some stable base for their measurements have in recent times been faced with the prospect of probably having to accept that their most cherished datum, known as the solar constant, is probably not constant. It was theoretically the fixed rate at which solar energy would be received by a unit area one metre square in an attitude normal to the direct solar rays placed just outside the earth's atmosphere, amounting to about 1.5 kW. And while the total energy input rate over the whole projected area of the earth in the direction of the sun is enormous, even the slightest variation is likely to change virtually all the elements of our climate.

It is somewhat amazing (providential or a fluke, whichever you prefer) that on the average, over the year, the reception of energy on earth is almost precisely balanced by the return loss to space. If this were not so a persistent net loss (or gain) would mean a continual change of average temperature, and over a relatively short time, life would not be as we know it, even if at all, without our current global helping hand. However, even though there is a near balance in the net solar energy budget, the energy we receive is not spread evenly over the surface of the earth.

The schoolroom globe again will show some of the reasons. Firstly, it will be clear, if we illuminate our model with a parallel beam of light, that some rays will strike the surface of the globe at right angles, others at varying angles down to a grazing

Fig. 3.4 Wind turbines, Sexbierum, Friesland (photograph by L. Delderfield).

incidence that is almost tangential. Therefore the intensity at the surface will also vary greatly, depending roughly on the latitude of the area we consider. Secondly, and partly because of this, the colours of the natural surface will vary from white to dark green. This is a direct expression of the amounts of visible light and other radiation absorbed or reflected.

Notice particularly that the polar regions, having received less intense energy in the first place, then carry a white surface which reflects back to space most of the energy they do receive. The reflective power of different surfaces is usually described by astronomers and climatologists as the albedo of those surfaces, usually expressed as a percentage of the radiation received.

Although there are many lists of figures to choose from, and it depends which wave-bands of energy are being considered, the following are typical:

	%
Pasture	19
Green fields	11
Conifer forest	7
Deciduous	9
Lakes/rivers	7 – 9
City areas	10
Sand	13 – 18
Clouds (average)	55
Snow	89 – 90

Thirdly, the areas of land and water are not distributed evenly over the whole globe. There is a much greater area of land relative to water in the northern hemisphere, and apart from different reflective values, land masses heat up and cool down more rapidly than masses of water.

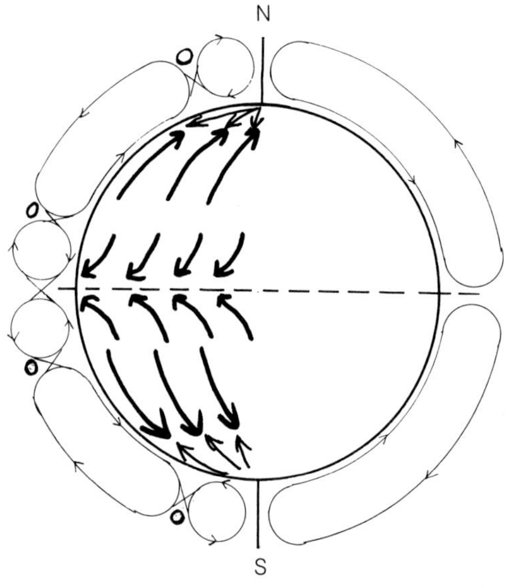

General circulation

All these variations mean that received energy is unevenly distributed over the earth, and a fluid medium — air or water — will tend to move to equalize this distribution. The general circulation of the atmosphere and, to a lesser extent, ocean currents, have been studied for centuries as being the sources of climate variations. The atmosphere is not much heated by the passage of radiant energy through it, rather it responds to contact with heated (or chilled) surfaces. It is noticeable that in the morning the ground heats up first and then passes heat to the lower layers of air. Conversely, on a clear evening, heat is lost rapidly from the ground, which then chills the air layer near to it: ground frost occurs generally before the air reaches freezing point.

Air heated by contact with a warm surface expands and becomes less dense. Being free to move it is then likely to rise and be replaced by cooler air in its vicinity. This is the basis of an elementary convection system known as a Hadley Cell after the 18th century physicist who proposed it as the basis of atmospheric circulation.

The actual pattern of circulation in the troposphere is much more complex than this; nevertheless, energy and other items such as water vapour and smoke particles are moved about in the lower atmosphere by such a mechanism. The accepted circulation pattern is now taken to consist partly of Hadley Cells, with intermediate motions in opposite directions, the whole thing rather like a celestial train of gears.

A large-scale Hadley Cell is compared, on the scale of the whole globe (though with the depth of the atmosphere vastly exaggerated) with what is, on the basis of recent research, more likely to be the system which actually operates. The additional factor which Hadley's explanation required was the earth's rotation; furthermore, the jet-streams, compact, high-speed airstreams, flowing a considerable height from west to east, were not known to exist until high-altitude flying became common during the 1940s. The flow patterns tend to interact like a train of gear wheels. The arrows show typical wind patterns as they are likely to be experienced near the surface; they are deflected from the paths one would expect on a stationary globe, to the right in the northern and to the left in the southern hemisphere.

Fig. 3.5 General circulation of the atmosphere.

A major secondary source of power which modifies the simple convection pattern is the rotation of the earth on its axis, which causes, for example, surface winds in the northern hemisphere to be deflected to the right (and to the left in the southern hemisphere).

Observations from high-altitude aircraft in World War II revealed other major flows, high-speed, relatively narrow jet streams, generally flowing from west to east. These streams have sometimes been held responsible for bringing North American winter weather to Western Europe.

Over the globe as a whole, the winter pattern will be different from the summer pattern because of changing day lengths. This roughly balances between hemispheres, and each continent and ocean will tend to have a recognizable, not to say always predictable, weather sequence.

For example, the British Isles have, very generally, two major weather expectations, since they are commonly either visited by warm moist air from the mid-Atlantic with frequent storms, or by cold drier air from northern Asia via Scandinavia. These two contending movements are there simultaneously for much of the time, and the polar front, which roughly separates them from south-west to north-east across Britain, wavers about for much of the year and produces much uncertainty among the population.

For example, it produces over a fairly small area of the world, a remarkable lack of confidence in weather forecasts and, perhaps, a lack of belief in all official pronouncements.

Climate change

The impression given in some quarters these days is that we should regard climatic change as a recent phenomenon. However, records by means of fossil plants and ice cores from periods over millions of years, suggest that change is the normal state. Major changes, such as glaciations and interglacial warming may turn out to be at least partly due to rhythmic changes in the earth's magnetic field. Superimposed on such long-period fluctuations are other more or less regular cyclic changes with periods of the order of centuries and decades. There is a more or less regular sun-spot cycle at about $11\frac{1}{2}$ year intervals. And sometimes you can in southern England have a very cold morning in May which seems like the onset of another glaciation.

Involuntary responses

We can reproduce in our imagination, or by experiment, some of the responses to environmental change or pressures. Consider first a sudden, severe environmental change such as falling unclothed into the North Atlantic in winter. Some automatic responses would almost immediately begin. Notably involuntary shivering, muscular spasms which would generate some heat, though at the expense of massive use of energy, to be cut short by lack of fuel and the onset of cramp; death would end the experiment in minutes. Other mechanisms might have come into action, such as vaso-constriction, i.e. the closing of small blood vessels near the extremities, effectively reducing the volume of circulation, economizing on blood-flow and heat loss, but this would favour the development of frost-bite and gangrene. A further response would be the emergence of goose-pimples, (pilo-erection) a rather pathetic attempt to raise the hairs on the body to entrap more air and hence improve the insulation layer.

Acclimatization

A considerably slower change of environmental conditions is usually experienced by the mountaineer. It is often supposed that bottled air or oxygen is essential on a climb which goes above 2500 metres. This is not so, provided that the ascent is made over a period of weeks. The process which operates here is acclimatization. It is even more successful if months or years can be devoted to it. There are tribes living at about 5000 metres above sea level in the Andes mountains of South America. They have changed physiologically, particularly in the ability of their blood to transport more oxygen, and in their pattern of breathing, which has become rapid and deep. The increased haemoglobin concentration, however, raises the viscosity of the blood, putting an extra load on the heart.

Techniques of climatic modification

Provided that there is a long time period, the human body can adapt to suit climatic conditions by means of genetic modifications. Whether the human species emerged from several centres simultaneously or diffused from a single centre, ethnic variations can be expressed in terms of five main racial types: Caucasoid, Negroid, Australoid, Mongoloid and American Indian. Skin colour does not greatly affect thermal comfort; it is more of an aid in filtering out harmful solar radiation than a way of increasing radiant heat loss from the body. The area:weight ratio of the body does affect radiant loss. The lanky build of the Bushman favours heat loss, while the squat build of the Eskimo favours heat retention.

Clothing and building are also techniques of climatic modification, and depending on the physical needs and materials available, have developed into traditional local forms. Before the comparatively recent transport revolution, it would have been possible not only to recognize the source of such vernacular solutions but also to deduce from them the climatic stresses or opportunities to which they were intended to respond.

Human intervention

The human species has continually influenced the natural climate, both purposefully and accidentally. Recent concern that we might be changing it comes rather late in the day. We have done this for hundreds of millennia. It is not too fanciful to regard lighting a fire in a cave as 'human intervention in climate'; the objective was to modify the weather (providing shelter and warmth), resulting in a device for improving the immediate climate, and to permit the colonization of otherwise hostile habitats. Clearing forests is also an ancient human practice also contributing to climatic change no less drastic by being accidental.

Just as there are scales at which we can consider the elements of climate, so there are scales at which we can recognize its modification. The climate in which we actually live exists at a very small personal (micro-climatic) scale inside our clothing.

The way in which we use our clothes and our buildings may be physical, ritual and social, but it is by the adoption of such personal climatic control systems that we have been able to survive severely unfriendly environments such as tropical deserts, frozen tundra, or the surface of the moon. The moon-walker survived because technology was available to enable him to carry

Fig. 3.6 Vapour trails from aircraft (photograph by A. Sealey).

around a flexible fabric structure (suit) which could trap a layer of earth-type climate next to his skin, and a servicing system (back-pack) which maintained this encapsulated climate — principally controlling air temperature and humidity. Back at his Lunar Exploration Module our moon-walker could shed his moonsuit and enter his communal overcoat or capusule. The astronaut's third skin had the same functions as a building; it had to be designed for a 'healthy' atmosphere, otherwise our astronaut would have soon died. The word design is the opposite of the word accident.

Deliberate change

Among deliberate climatic changes perhaps the most popular in recent decades has been rain-making. This was once the province of the witch-doctor, and we still keep our fingers crossed during long periods of drought.

Italian wine growers fire explosive rockets into thunder clouds just as the grapes are reaching ripeness. A similar principle is the Stiger Vortex Gun, an Australian device which directs shock waves towards the clouds from a static mortar on the ground.

Although the theory of cloud seeding is complex, since the 1960s it has become a flourishing business in the USA, providing employment for both aviators and lawyers, who find a profitable field of actions for theft of rightful rain, as well as imposition of unwanted rain. The commonest procedure for cloud seeding is to inject into pregnant looking clouds either a 'smoke' of finely divided silver iodide crystals, dry ice or salt to act as nuclei for the formation of raindrops. Such operations can only be described as creating climatic changes if carried out frequently, so as to represent something more than a small-scale weather change. Current annual heat waves and lack of rain in many parts of the world could possibly be more than the natural order of things.

Selfish human actions causing climatic change in the face of scientifically based warnings must be considered criminal.

Shelter belts

A well-established and successful technique of constructively changing local climate is used by foresters anxious to protect stands of young trees in their most vulnerable stages of growth, as well as by farmers to shelter tender crops.

Such shelter belts are contrived of several graded rows of quick growing shrubs and trees to reduce wind speed near ground level in the area to be sheltered. The mechanism is based on the increase of small-scale turbulence, for this is the only way to take speed out of wind. A solid barrier might take up less room but would be less effective since the deflected airstream would attack the ground with more severity than the free wind some distance downstream.

Equally, a single row of trees cannot form a shelter belt, since, while it will cause some turbulence, the airspeed near the ground will be increased in the gaps between the trunks. A combination of single rows of trees (such as poplar) interspaced with a quick-growing hedge of hawthorn and open-weave matting can result in economic use of minimal space. (The same problems exist with the use of tree and hedge divisions between busy roads and buildings where modification of man-made polluting micro-climates occurs. See Chapter 4, p. 50.)

Fig. 3.7 Trees and a hedge combine to form a wind screen to a hop yard in Herefordshire (photograph by A. Sealey).

Origin of the atmosphere_____

Trees are climatically important in other ways, and, on a major scale:

some living cells can live in an atmosphere without oxygen. Some such organisms may have been the first living things on earth, preparing the way for the earliest green plants, which in turn would supply most of the free oxygen to support higher forms of life.[5]

Green plants represent an important ecosystem, and illustrate well the concept of ecology — a popular word which strictly means 'the study of the household'. The word 'household' is here taken to mean all living things in relation to their surroundings, and an overall principle of the subject is that it is virtually impossible to change one element of the natural world without affecting many others. Though the so-called 'balance of nature' was never static we tend to expect it to favour human survival although there is no reason to suppose that it should.

Indeed, the human species has been on this earth for a much shorter time than the dinosaurs survived, whom we often think of as inefficient. It is a tragedy that among the results of advanced technology we have now the ability to destroy our natural environment at far greater speed than ever before. The depletion of forests, which began with a chipped flint hand-axe, has led to changes in water levels, massive erosion of soil and loss of plant and animal life, as habitats are violated due to ignorance, greed, or just thoughtlessness without malice. But sustainable growth of forests, with good housekeeping, can supply us with a building material to suit many of our needs.

Accidental change_____

A whole group of accidental weather changes has occurred over the centuries, some brought about by the fallacy that you can dispose of unwanted rubbish by burning it, when burning only changes the form and location of the pollution, and may exacerbate the problem. This was graphically illustrated in the late 1960s when a large laden oil tanker ran aground off the Scilly Isles and began to discharge its cargo. The magnitude of the spillage made it apparently impossible to contain or collect the oil. To prevent the beaches becoming coated it was decided to burn the oil away. The Air Force, appreciating bombing practice, set the oil alight, resulting in a spectacular cloud of smoke and visible water vapour with considerable vertical development due to thermal convection, followed by heavy rain and thunder (the Hiroshima syndrome).

Less obviously dramatic, but producing longer-term weather changes, are the cloud-making activities of cooling towers (mainly clean vapour), causing above-average rainfall downwind, and factory chimneys emitting and causing — who knows what? Building the stacks higher only extends the pollution, with plumes overlapping and advancing on a wider front.

Since the 1950s there has been a vast increase in air travel over Western Europe and the Atlantic Ocean, whether by piston-engined aircraft or jets: the pollution effect has been roughly the same, carbon dioxide and water vapour. It was observed that cloudiness over the area increased, and anyone who lived under a regular flight path would notice how quickly a clear morning sky could be veiled in thin cloud and eventually become overcast, frequently with contrails.

In the early days it was claimed that there was no real problem; both gases were said to be harmless. This was false. Both are now recognized as 'greenhouse gases'. If they lie around in the

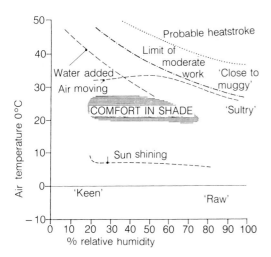

Fig. 3.8 Olgyay's bioclimatic comfort chart takes air temperature and relative humidity as basic elements of human thermal comfort, with the possibility of extending the comfort zone due to the presence of solar radiation or cooling breezes. Subjective impressions have been identified approximately on the chart (Olgyay, Victor, *Design with climate*, © 1963 by Princeton University Press. Reproduced by permission of Princeton University Press).

troposphere they may become part of the one-way celestial valve which lets short-wave solar energy through but impedes the slightly longer-wave return radiation trying to get out, leading to a net thermal gain near the earth.

Concern about cleaner air in the UK and other Western European countries from the 1950s onwards certainly led to dramatic reductions in solid particle deposits from industrial and domestic emissions and the reduced incidence of smog. However, it did little to reduce other less visible pollutants which give rise to acidic rain. It should not be forgotten that rainwater is often naturally acidic, particularly if it originates from thunderstorms, and that a certain nitrogenous content has traditionally been regarded as agriculturally useful. We are now facing a new concern, which extends to those Eastern European countries who stayed strangely quiet about their own environmental pollution activities in a blind race for industrial equality with the West.

Comfort conditions

Climate is not the only source of external influence on comfort, but it has always been significant. The term 'comfort' can be defined in a number of ways, but if we refer back to the previous comments on survival, then there are several factors intimately connected with elements of climate and their relationship to the human organism. We may casually use such phrases as 'keeping warm' or 'cooling off' which actually refer to finely balanced thermal problems. The inner organs of the body are very sensitive to temperature change, for instance the brain and some important glands depend for their correct functioning on temperature being maintained within very narrow limits, perhaps within a fraction of a degree, while the extremities of the body can tolerate a much wider range; the skin temperature can vary by 5−10°C over the body with little discomfort, whereas the doctor will expect mouth temperature to remain near 37°C to be regarded as normal.

While it is temperature that the doctor often measures first, it is the actual rate of heat loss or gain that is the real concern. In the course of absorbing and digesting food the body produces waste heat. It also gains heat from and loses heat to its surroundings by various mechanisms, and balance is achieved by the interaction of all these processes.

Numerous attempts have been made to construct a comfort index which could, in the form of a single number, assess the probability of achieving total comfort under stated conditions, or which could perhaps define the modifications necessary to be applied to the imposed conditions in order to achieve comfort. In his book *Design with Climate — A Bioclimatic Approach to Architectural Regionalism*,[6] Victor Olgyay took air temperature and percentage humidity to be the basic components of thermal balance. He postulated that a range of temperatures of roughly 20−27°C and a humidity range of about 20−80% could indicate at least the probability of thermal comfort being achieved '... by a lightly clad American male, resting in the shade ...' (truly an idyllic picture).

Human thermal balance

The comfort zone indicated on Olgyay's chart can be extended; it is common knowledge that comfort can be achieved, even without artificial aids, in conditions of lower air temperature, provided, for example, that radiant energy is simultaneously available (e.g. sunbathing in a snow field). Similarly, in

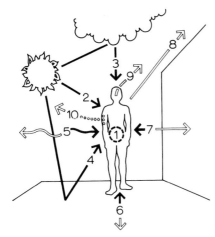

Heat gains ━━━━━ Heat losses ═══════
1 Body processes
2 Direct solar radiation
3 Reflected radiation
   from the sky
4 Reflected radiation
   from surfaces
5 Conduction from
   warmer air (assisted
   by airflow and
   convection)
6 Conduction from
   warmer surfaces
7 Radiation from
   surfaces

5 Conduction to cooler
   air (assisted by
   airflow and
   convection)
6 Conduction to cooler
   surfaces
7 Radiation to surfaces
8 Radiation to space
9 Respiration
10 Perspiration

Fig. 3.9 Components of thermal balance
(after Olgyay).

considerably higher temperatures, as long as the air is moving so that excess heat can be rapidly removed. There is, however, a practical limit to the higher temperature which can be tolerated, for if the air is above 33°C (common skin temperature) we will have to speak of a warming wind, rather than a cooling breeze.

It is interesting that Olgyay's selection of two basic parameters and two modifying variables reinforces the idea that we must be prepared for elements of weather to operate simultaneously, not singly. It also enables us to put more or less subjective descriptions to measurable factors. For example, low temperature and low humidity acting together (keen conditions) are usually more acceptable than low temperature with high humidity (raw). It is notable, too, that Olgyay's comfort zone cuts off steeply when external conditions are hot and humid (close and muggy) since in such a situation the possibility of offsetting the high temperature by evaporative heat loss is reduced if the moisture from the skin cannot evaporate readily. This problem used to arise frequently in some industrial processes, such as when unloading and recharging brick kilns that were still hot, when the difficulty of losing excess body heat during heavy exertion could lead to heat exhaustion and collapse. Although such industrial tasks have largely disappeared in Western Europe and the USA, they are still very much in evidence in China, India, Russia and Eastern Europe.

The building as a climate filter

It has been said that 'everybody talks about the weather; nobody does anything about it'. The human animal started to do something when he lit a fire in a cave. But there is more to architecture than keeping the rain out. Architecture was once defined as 'shelter plus'. One major stimulus to build has been the need to contain an acceptable total climate, including sound, smell and privacy as important elements, when the outdoor climate, as offered, is not acceptable or suitable for the chosen activity. Incidentally, the distance of external measurable conditions from the boundary of the comfort zone indicates the work that has to be done by the fabric and the servicing system.

With the cocktail of man-made pollutants now upsetting the normal natural way of things, the need for the building to act as a clean climatic filter becomes evident, as this primer guide shows. Thus the parts of the fabric of a building have a far more complex set of functions to perform than merely the delineation of space. The window may have to permit the flow of natural light (including vision), fresh air (at times) while also impeding the passage of excessive solar heat, driving rain, snow, unwanted sound and intruders. At the same time it is a primary facial feature by which we recognize the building and make visual judgements about it.

Therefore the total building fabric is a set of controllable climate filters, which can be handled or programmed to maintain an appropriate endoclimate against a wide range of variable constraints including outdoor and indoor pressures, such as are found in a hospital where people have different needs yet are all contained in the same space. Modular micro-climatic internal conditioning is a possible solution.

Energetics of climate

There are long-standing traditions that some regions offer climatic conditions which are conducive to hard work, where others promote lethargy. Readers from other lands will have their

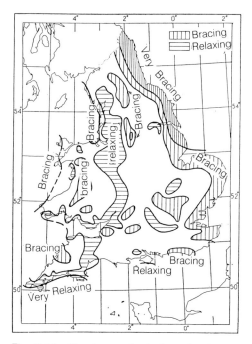

Fig. 3.10 Bracing and relaxing climates, England and Wales. (From Brooks, C.E.P. (1954) *The English Climate*, EUP.)

own examples. In England the North Sea east coast town of Skegness is bracing; whereas the University City of Oxford is relaxing.

A map of climate or climatic energy resembles a map of progress far more closely than does a map of any other factor. Under certain optimum conditions man's physical and mental capacities are at a maximum, his power to work the greatest, his initiative the highest and his ability to resist disease correspondingly high. Any departure from these conditions means less efficiency, mentally and physically, poorer health and a higher death rate.

E. Huntingdon *The Human Habitat*[7]

Huntingdon proposed the following essential factors for the best climate for human health and activity.

1 A fairly strong but not extreme contrast between summer and winter when the summer temperature averages not much higher than 19°C for day and night together.
2 There must be rain at all seasons, enough for the air to be moderately moist much of the time.
3 Constant but not undue variability of weather is almost as important as the right conditions of temperature and humidity.

The failure to appreciate the great importance of variability in the weather is one of the main reasons why the pervading effect of climate and changes is even yet only dimly appreciated.

Urban climate

While there may be many different opinions about what the building designer ought to do, there can be little doubt that one of the things the architect, together with engineers, planners, builders, occupiers and building developers should not do is to cause physical harm to the erector, the user and the passer-by. After many centuries of constructing habitable buildings there have emerged in recent years new ways of harming people.

Not that the sick building syndrome is a new problem. Marcus Vitruvius Pollio, a Roman engineer and architect in the first century AD noted:

The town being fortified, the next step is the apportionment of house lots within the walls and the laying out of the streets and alleys with regard to climatic conditions. They will be properly laid out if foresight is employed to exclude the winds from alleys. Cold winds are disagreeable, hot winds enervating, moist winds unhealthy. . . . Mytilene is a town built with magnificence and good taste, but its position lacks foresight. In that community when the wind is south, the people fall ill; when it is north-west it sets them coughing; with the north wind they do indeed recover, but cannot stand about in the alleys and streets owing to severe cold.[8]

Siting has not by any means always permitted free choice; there have traditionally been many factors besides purely climatic ones, e.g. fortifications, water supply, industrial zones, politics. But it still may be possible to avoid the worst sites provided we can re-educate ourselves in the skills of nest building which we once had.

Topoclimate site selection/orientation

Any new site already modifies the climate of the region; for example, topography will affect day length, water courses, airflow, the frequency of frost and fog, and rainfall patterns, not forgetting other factors which are set out in Chapter 4. It would be possible to appraise a potential building site, even without weather records, by reference to a contoured survey map assisted by local knowledge. The local inhabitants (e.g. farmers) tend to know about the local micro-climate; soft-fruit growers will be aware of the location of frost hollows in which on clear calm nights, when the air is locally cooled by contact with the ground which has lost heat by radiation, the air becomes more dense and flows to lower levels. It may collect in pools of very cold air or fog.

Frost hollows can also be man made by the building of a railway, or road embankment across a valley, thereby blocking the natural, down-slope (katabatic) flow of chilling air. Farmers seem to have been aware of this long before planners.

There is a classic case of feed-back which develops between a building and its site; if we build at least partly in order to create

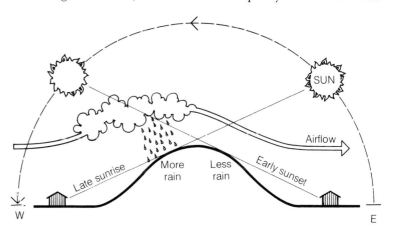

Fig. 3.11 Section through N–S hill range showing typical local climatic conditions. Drawing by A. Sealey.

a suitable endoclimate and maintain it with the minimum of artificial aids — an economic and aesthetic object — then the fabric of the building should be designed in form, composition and detail to deal with those outdoor conditions which impinge on it (see examples in Chapter 5).

But, the designer having built for such reasons, the outdoor conditions which remain will themselves be modified by the

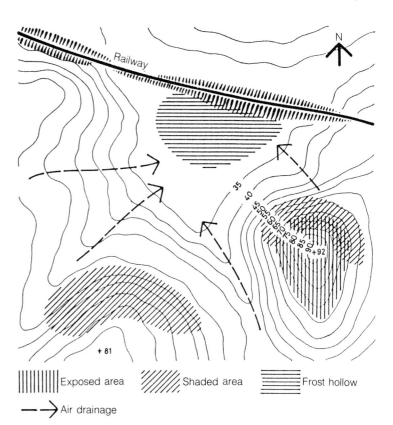

Fig. 3.12 Topography of a moderator of climate: since local topography will impose variations on regional climate, it is possible to appraise some of the likely qualities of a site by reading a contour map. Drawing by A. Sealey.

Fig. 3.13 Fog, Worcestershire (photograph by A. Sealey).

simple presence of the built form. Once there is a building, or even a garden wall, it has an exposed face and a sheltered face, and the spaces around the building are then subjected to a new climate, whether natural or man made.

It would be unrealistic to suggest that the designer has total freedom of choice in the matter of siting a building, but the greatest test of professionalism is when to advise that a particular building should not be built, or that its siting should be subject to an interaction (or synergy) of many diverse factors (see Chapter 5).

Individual buildings and whole new settlements have been badly sited during the past 2000 years. When a virgin site is described as 'ripe for development' in a region where building has happened over thousands of years we should ask why the site has not been used before. The answer may well be climatic, which of course includes the nature of the land on which the building is proposed to be built.

Aesthetic and practical issues have always been connected; there is no truly permanent building material, and replacement or maintenance costs are recurrent. Furthermore, the expense in energy necessary to maintain comfort conditions is also now well understood. An understanding of comfort conditions as they relate to indoor air quality is leading to a new appreciation of the use of low-energy natural materials and the ability to modify and use the natural climate to eliminate as far as possible the use of climate-polluting fuels.

Building aerodynamics

It was somewhat surprising in view of the long history of building tall, in the USA for example, that new problems emerged in the late 1940s, when it was found that a slab building did not behave in the same way as a tower. A slab offers a large sail area to the wind and virtually all the air that hits it is deflected around it, including over the top. The first vertical slab of considerable size was the United Nations Secretariat, built in 1948. Soon after completion it was inevitably subjected to heavy rain, some of which ran upwards over the surface, which it was not expected to do. The weathering system of curtain walling had been designed on the assumption that water runs down a wall, and so the windows leaked upwards. To perform against horizontal rain they should have been designed as though for an aircraft, which is virtually what a tall building is for design purposes.

Did architects and planners in the 1960s know or care about

Fig. 3.14   Urban texture — UN Building, New York (photograph by Bill Holdsworth).

[The standard work on air-flow problems, *Wind Environment around Buildings*,[9] includes an approximate relationship between urban airspeed, safety and comfort.

Onset of discomfort	5 m/sec
Unpleasant	10 m/sec
Dangerous	20 m/sec]

the aerodynamics of buildings? How many had visited New York City and learnt from bitter experience the funnelling and downward effects of the wind? Was it thought only as a bridge builders' problem? Direct model analysis, always attractive since you can see what can happen, was a well-established tool in the field of aircraft design. It is recorded that Gustave Eiffel had studied a model of one of his lattice arch bridges in much the same way as bridge analysis was undertaken later. But most of the wind tunnels available in the mid-20th century were made to provide for high-speed structural testing in a steady flow, whereas with tall buildings and factory chimneys the problem was often to do with environmental nuisance caused by relatively low-speed flow in a special situation where horizontal wind-shear is present (i.e. in the lower levels of the troposphere where the airspeed varies with height above ground due to friction at the earth's surface).

At the time of writing, curtain walling has gone out of fashion. Tall buildings continue to be erected, and techniques to reduce air infiltration have resulted in internal environments being sealed. The consequence has been problems, often unforeseen, of a polluting internal micro-climate, and people are now calling for fresh air. Whatever the method of building, the effect on the natural environment remains, and the real art is to achieve a healthy balance.

New climates and resources

Early in the 19th century the meteorologist, Luke Howard (who gave the clouds their specific names), predicted in what ways the climate of a city would be different from that of the country around it. In 1965, it was confirmed by surveying London from a travelling weather station that a 'heat island' was a common characteristic of a densely built-up urban area. Any building is likely to be a source of heat, either wasted from heating or cooling plant, or stored solar energy released from the walls after sunset. It was found not unusual for a difference of 6°C to be measured from outer to inner city regions. It would be interesting to determine what effect the increase of traffic and the emissions of lethal fumes have on the temperature increase in the current situation.

While the view from the penthouse flat may be superb, the top of the tall block will sometimes be above cloud base, and while it could help to reduce the severity of ground-level airflow if the surface of tall buildings were modelled in deep relief, the best solution would be not to build any city block more than twice as high as older buildings in the area. This has not stopped the building of Canary Wharf tower blocks in London's old dockland. All buildings, even if underground (for they too are likely to emit waste energy), represent an intrusion into the biosphere. The winning and processing of materials always modifies the landscape, even if we discover in time the massive advantage of that superb stuff called timber, the only renewable building material. Why bother to keep a cow, as the man said; if you know the chemistry of all the dairy products you can synthesize them without all this business of growing fodder and disposing of the waste. Whatever the disadvantages, the cow is a compact, mobile, solar-powered protein factory. The tree is a solar-powered factory producing among other things a structural material about as efficient — strength for weight — as mild steel. It is clearly a monumental task to replant the world's forests at anything like the rate at which they have and still are being destroyed.

Fig. 3.15   'The new Manhattan' — Canary Wharf, Isle of Dogs, London (photograph by Simon Holdsworth).

There are some signs that the human race is gradually improving its attitude to waste, whether in the form of reusable materials or recoverable energy. In affluent societies where water is plentiful, it has become accepted that water is used lavishly and then discarded. Even relatively clean rainwater is emptied into foul-water sewers; this may keep the foul drains cleansed but makes the water harder to recover. In the early 20th century the English country house was equipped in the service area with a water tap to enable the foundations to be watered. The intention, especially on shrinkable clay subsoils, was to reduce the risk of subsidence likely to occur during unusually dry summers.

Reference is made in this book to the work of Dutch, Norwegian and other European architects and to techniques that were common in building construction 100 years ago. Lime for mortar, and bricks from suitable local clay produced locally, enable employment to be decentralized, and that production could be more friendly to the local atmosphere than bulk production methods. It can also be argued that rain which falls on city conurbations as well as on individual dwellings should be collected; it could remain on the site, collected in small lakes. Ponds and lakes within a built area are a useful moderator of temperature, both lowering high temperatures and acting as a source of stored heat energy. Many such examples of the former are found in Sweden, and of the latter in the Netherlands.

With the growing water shortage, due in part to global climatic changes, houses and buildings should be provided with water collection stores as a standard amenity. If high-quality buildings meant for healthy habitation cannot function in physical terms without prohibitively expensive engineering services, then the designer has not fully understood the age we live in. Solutions must be both economic and sustainable.

Architecture, a fine and useful art, cannot flourish in the absence of an effective technical basis; and this, in turn, cannot be appropriate or efficient in ignorance of the natural conditions against or with which it is expected to function. Therefore a knowledge of all external environmental factors, whether hazards or opportunities, must be the concern of the designer. Building effectively is the necessary starting point for healthy living.

It was Rayner Banham in *The Architecture of the Well-tempered Environment*[11] who suggested that if you were cast up on a remote island with a limited supply of timber and were faced with the prospect of a cold winter, you would have to decide quite quickly how much to burn, and how much to use to build a shelter. Both decisions involve deliberate climatic change. If this should happen to the whole human race we would soon regret not having replanted the trees that we had destroyed earlier.

If Britain were left to itself, as an abandoned site, it would quickly revert to forest, dominated by sycamore, birch, ash and oak. If, rather than by long cold winters, we were threatened by moist mild winters and long hot summers, we should still need shelter, with the trees and plants that can also provide fuel and food growing even more quickly.

This book is written to encourage you to survive, and to learn something useful towards that survival. The chief maxim for all prime movers, designers, constructors and users is that we should not be surprised by the results of decisions made.

4 Factors of influence

ECHOES: Environmentally controlled human operational external-enclosed space: a planning guide

In the creation of a well-tempered internal environment an environmental design matrix that correlates human requirements and the physical sciences helps us to achieve our wants and needs.

The 18 different sections shown in Table 4.1 can interact in many different ways. Although sunlight is referred to under luminous/passive control, it is the only item that seems to have an immediate connection to natural climate in the table. The other, more obscure, reference is contained within the phrase 'the thermal environment'.

This was the traditional approach to internal building climate design. Solar gains and heat loss calculations were based on generalized weather information that often resulted in large variations between actual and design code minima and maxima for temperature and humidity, with resultant inefficiencies in fuel use and building design methods.

The traditional approach to design is one of the many reasons why buildings have become 'tight'. They were designed with the minimal amount of outside air allowed to infiltrate into the building, and air quality control was regulated to laboratory norms that simply did not suit the real world.

*Source Herbert J Spinden *Songs of the Tewa*. New York Exposition of Indian Tribal Arts, 1933

Table 4.1   Environmental design matrix: creating a well-tempered environment

	01 Spatial factors	02 Luminous factors	03 Sonic factors	04 Thermal factors	05 Matter and energy	06 Correlation
1 Man–environment relations	Anthropometrics Proxemics Space requirements	Light Psycho-optics The luminous environment	Sound Psychoacoustics The sonic environment	Heat Psychothermics The thermal environment	Energy and its conversion Needs and waste provisions	Psychophysics Human — ecology (man — environment systems)
2 Passive controls	Space Organization (planning — movement)	Daylighting (including sunlight)	Unwanted and wanted sound (noise control and room acoustics)	Passive thermal controls (fabric and form)	Hygiene (water supply, drainage, waste disposal)	Integration in building terms: (the built environment)
3 Active controls	Communications Movement Protection	Electric lighting	Electroacoustics	Heating Ventilation Refrigeration Air-conditioning	Energy systems (electricity, gas, oil, vacuum, compressed air)	Integration in performance and hardware economy

Taking into account all the outside factors, then moving back into the conceived space, and then returning to the outside, created a circle of activity, a wholeness. In undertaking this design journey we are now beginning to find many other man-made factors of influence which are unnatural to the attainment of an environmentally controlled human operational external and enclosed space.

To undertake this design journey we shall in this chapter divide up the areas of proposed study. Not every section is dealt with in detail or at all. All designers have at their fingertips design guides and reference codes. This book will concentrate on the less known and less accepted factors of influence.

1 External
   (a) Climate: meso- and micro-assessment: solar factors, wind and rain
   (b) Composition of air: air quality
   (c) Composition of soil: ground-water/pollution
   (d) Radiation in both air and soil
   (e) Noise and vibration
   (f) Relationship to other buildings: the surrounding envelope, shade and openness
   (g) Vegetation and visual appetizers
2 Internal
   (a) Construction physics: thermal comfort, surface temperatures of walls, air, air velocity (ventilation), humidity; illumination and daylighting; composition of air; radiation; noise and vibration, sound levels
   (b) Building materials: volatile emissions (outgassing)
   (c) People's needs: working circumstances to relate to operational pollutors; furnishing materials for biological or chemical contaminants; waste products; colour
   (d) Technical installations: indoor air quality, sound, moisture, electrical services, telecommunications, fire; lighting
   (e) Occupational behaviour: direct influences (e.g. cooking, washing, corporal excretion, furnishing); influence of building services and equipment
   (f) Vegetation and visual appetizers

To complete the circle, there are the town planning requirements of roads, walking, cycling and other transportation, the collection of waste and disposal, and the effect of the building in relation to other elements of the built space.

Fig. 4.1 Liverpool front from River Mersey, 1980 (photograph by Bill Holdsworth).

In January 1990 a proposal from a working party of the Technical and Quality Control Department of the Rijksgeouw-dienst (National Building Institute), The Hague, The Netherlands, also included in a similar matrix for phased investigation three stages, namely inventory, parameters and demands.

Our intention is to introduce to the designer/decision-maker information on many of the new parameters that are daily coming into the design process for buildings. Emphasis has been placed on some subjects that are still new, and possibly not fully understood, such as electromagnetic fields of radiation, the effects from outgassing, the emission of toxic particles from building materials, and the resulting risks to human comfort associated with the damage we are doing to our natural climate.

Some subjects are still treated with disdain and disbelief by many people long in the profession of designing buildings. Although I believe that such conservatism does not help to solve our problems, where information has proved difficult to check I refer you to the Resources Network, pp. 136—7.

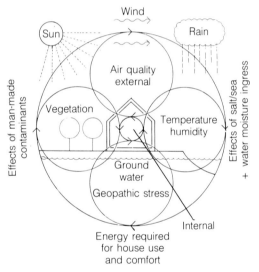

Fig. 4.2   Factors of influence (after Okerkamp, 1988).

4.1   Factors of influence: external

Meso- and micro-climatic assessment_____

The raw material of climatology, such as atmospheric pressure, precipitation, temperature, humidity, radiation and air movement data is supplied by the science of meteorology.

Climate can be defined as an integration in time of the physical state of the atmospheric environment at any particular place. To be able to attempt any such integration we first examined in Chapter 3 the factors shaping the natural climate. We shall now look at the methods of observation and of handling enormous amounts of data.

Radiation, spectrum

The earth receives almost all its energy from the sun in the form of radiation. The spectrum of solar radiation extends from 290 nm to about 2300 nm:

1 Ultraviolet radiation, 290—380 nm, produces photochemical effects, bleaching, sunburn
2 Visible radiation, 380 (violet)—700 (red), perceived as light, but also produces some heat effect
3 Short infra-red, 700—2300 nm, mainly perceived as radiant heat, but with some photochemical effects
   (1 nanometre (nm) = 10^{-9}m)

Radiation quantity

The amount of radiation reaching the earth's surface depends on the following factors:

1 The rate of energy flow at the upper limits of atmosphere (the agreed mean value of this solar constant is 1395 W/m^2 with a variable of $\pm 2\%$ due to sunspot activity, and $\pm 3.5\%$ due to variables in solar distance

Fig. 4.3 The flowing contours of the NMB Bank, Amsterdam (photograph by T. Alberts).

Solar constant = 100%

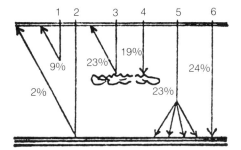

1. Scattered into space
2. Reflected by earth
3. Reflected by clouds
4. Absorbed in atmosphere
5. Diffuse: absorbed
6. Direct: absorbed

Total on surface = 5 + 6 = 47%

Fig. 4.4 Radiation quantity.

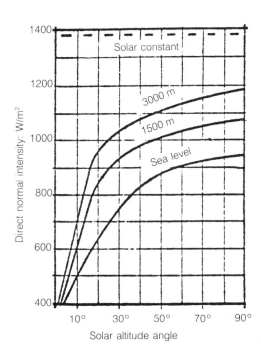

Fig. 4.5 Solar altitude angle.

2 Atmospheric depletion, depending on the transparency of the atmosphere such as vapours, ozone depletion and dust
3 Solar altitude angle (length of a path through the atmosphere, depending on geographical latitude, time of year, and hour of the day), the altitude above sea level, and the angle of incidence

The daily amount of radiation also depends on the length of the sunshine period, which is a function of the geographical latitude and the time of year.

Wind generation

The maximum radiation is received on the earth's surface at the latitude circle of the sun's zenith path which moves between the two tropics of Cancer and Capricorn. The same movement governs the world's wind patterns.

Of particular significance to our design criteria is the way in which winds move local to our surroundings. One local air movement of special interest is the katabatic wind. Particularly on clear nights, due to intensive outgoing radiation (to space), the ground surface can cool far below the temperature of the air. The layer of air in contact with the ground also cools down and is accompanied by the formation of fog. When this cold air settles in depressions it behaves as a liquid: it can flow down a slope, or in a valley, it can fill other depressions, or it can overflow certain barriers. These frost hollows were referred to in Chapter 3 and together with small-scale convective wind generation can be important factors in changing the immediate micro-climate.

Wind data

While meteorological stations record wind directions and velocities daily, continuous recordings can be made by the use of an anemograph. For quick reference the most useful form of data presentation is the wind frequency chart.

Temperature

Temperature conditions are governed by radiation and wind. An area strongly heated by the sun will be warm, but the intrusion of a cold air mass can completely change the conditions.

The solar heating effect itself is a function of the transparency of the atmosphere; for example: cloud cover. On the other hand, strong heating accelerates evaporation, and increases the moisture content of air, which at a higher temperature can support more moisture. As the air moves to a cooler zone, its temperature drops, vapours condense, and clouds are formed. It is a highly complex pattern with many variables. Records show that in Libya in 1922 the maximum was 58°C, with a minimum of −86°C recorded in Antarctica in 1958.

Precipitation and humidity

Although the amount of precipitation, the collective term for rain, snow, hail, dew and frost, does not directly influence atmospheric comfort conditions, it does affect the type of vegetation of a location and governs the ecological balance. Comfort feeling is influenced indirectly or through the associated humidity conditions.

Fig. 4.6 Urban landscape — Glasgow, Scotland (photograph by Bill Holdsworth).

Rain penetration results from a combination of intense rain and wind. This factor is illustrated in Chapter 2, p. 6, where the rain also brought corrosive salt from the Atlantic Ocean (some several hundred miles away) to the conurbation of Glasgow: a factor that no builder worth his salt should forget.

Humidity is almost exclusively measured as relative humidity in percentage terms (r.h. %). Maximum humidities are associated with the lowest temperatures of the day, which usually occur just before sunrise; minimum values are expected at the warmest times of day, early afternoons being more characteristic for a given location.

Locations will show diurnal and annual variations for radiation, temperature and humidity. In the past designers have not normally taken much interest in the great mass of climatic data available. It has tended to be a variable that we believed we could ignore. In doing so we have failed to see ourselves as part of a living world.

The Preseli hill range in western Wales rises some 200 m above sea level; the land is swept by winds and rain from the open sea some few miles away. Yet the natural timing of spring blossom and growth is often several weeks behind that in the protected hidden valleys lower down towards the sheltered bays.

An architect friend recently wondered what climatic change would be experienced if a sustainable roof garden were built at the top of the giant skyscraper looming up over the Isle of Dogs at Canary Wharf. In return I wondered whether the problems of indoor air quality control were different from the bottom of the building to the top. Without any scientific evidence I would believe that they are, and therefore would propose in any design team planning session that a detailed evaluation would be useful.

Apart from questions of shelter, protection, the provision of fresh water, and routes for trade, one of the most important items of natural climate for people was humidity. Ancient cities were found to be built on what has become known as the 21°C line. But it was also where the humidity allowed people to be comfortable.

Fig. 4.7 Cumberland Wharf, Docklands, London. A good example of a building designed to use wind/daylight and solar energy to maximum advantage. Architect: Alan Turner RIBA (photograph by Bill Holdsworth).

1 Twelve vertical lines are drawn, positioned to show the 15.00 h humidities, extending vertically to give the monthly mean range of temperatures.

2 The position of the area defined by these lines, in relation to the comfort zone, immediately shows the basic nature of any discomfort condition.

3 The cooling effect of winds can now be considered. In this case, the given conditions being largely below the comfort zone, there is no need for cooling. Winds can be excluded, velocities not being very high, thus need not be considered any further.

4 Radiation effects are quite significant. The coolest and the warmest months will be checked.

July is the coolest month, the mean maximum just reaching the lower comfort limit. In this month the daily average radiation amounts to 14.4 MJ/m^2 day. For 12 hours of daylight (and 3600 seconds in every hour) this would give an average intensity of

$$\frac{14\,400\,000}{12 \cdot 3600} = \frac{144\,000}{432} = 333 \text{ W/m}^2$$

This amount of radiation would lower the discomfort limits by 333/73 = 4.5°C (also apparent from the chart). With this, about half of the July line would be within the comfort zone. The upper limit is not reached. In the absence of radiation, at night, cool discomfort is likely to prevail.

March is the warmest month. Temperatures range from comfortable to cool discomfort conditions. Radiation is 23.4 MJ/m^2 day.

$$\frac{23\,400\,000}{12 \cdot 3600} = \frac{234\,000}{432} = 541 \text{ W/m}^2$$

This would lower the comfort limits by 541/73 = 7.4°C. It is apparent from the chart that temperatures will be above the lower comfort threshold and at some times the upper limit may be exceeded.

Conclusion: a very amicable climate. Overheating can only be caused by radiation. Particularly during the warmest months, some form of solar protection will be necessary. Cool discomfort occurs at night in all months, especially in July and August.

Fig. 4.8   Climatic graph for Nairobi (after S.V. Szokolay).

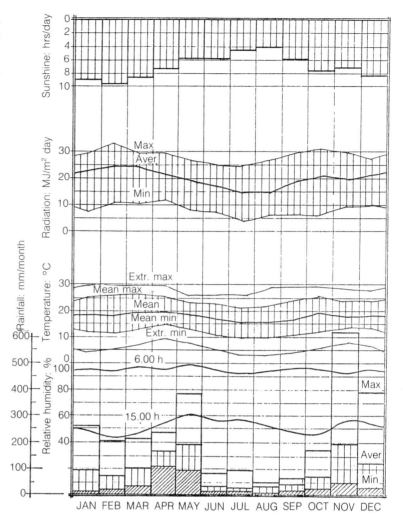

Climatic graph

An important tool is the use of a climatic graph. The example given is based on data for Nairobi, 01 17 South, 36 48 East.

It is important to group the large number of variables into a number of climatic types. There are several types of climatic classification in use, most of them based on dominating factors:

1 Climates dominated by equatorial and tropical air masses
  (a) Rainy tropics
  (b) Monsoon tropics
  (c) Wet and dry tropics
  (d) Semi-arid tropics
  (e) Arid tropics

2 Climates dominated by both tropical and polar air masses
  (a) Dry summer subtropics
  (b) Humid subtropics
  (c) Marine climate
  (d) Mid-latitude arid climate
  (e) Mid-latitude semi-arid climate
  (f) Humid, continental warm-summer climate
  (g) Humid, continental cool-summer climate

3 Climates dominated by polar air masses
  (a) Taiga
  (b) Tundra
  (c) Polar climate

4 Climates having altitude as a dominant factor
  (a) Highland climate

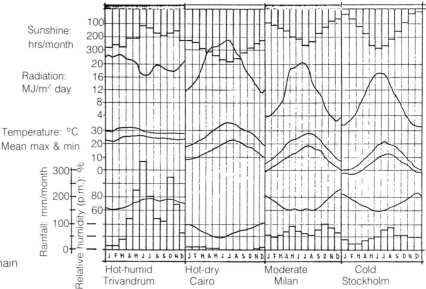

Sunshine:
hrs/month

Radiation:
MJ/m² day

Temperature: °C
Mean max & min

Rainfall: mm/month

Relative humidity (p.m.): %

Hot-humid
Trivandrum

Hot-dry
Cairo

Moderate
Milan

Cold
Stockholm

Fig. 4.9 Climatic graph for the four main climatic areas (after S.V. Szokolay).

For architectural design purposes these different classifications are further broken down into four main areas of study: hot-humid, hot-dry, moderate, and cold climates.

Hot-humid climates   In hot-humid climates there is very small variation throughout the year. An example is the Israeli city of Tel Aviv where the mean summer daily temperature is 35°C with a mean r.h. of 82%. Winter temperatures are usually not less than 20°C.

Hot-dry climates   These can be defined as having high radiation values and high temperatures. The mean temperature of the warmest months is well over 25°C with low relative humidities. Maximum and minimum temperatures in a year can vary from 45°C to −10°C; winds are strong, not being restricted by vegetation, and carry much sand and dust. But as I was to find out from living in such a climate, Cairo should be a healthier place to live in than Tel Aviv. In natural climatic terms it should be, but because of a lack of good sanitation, poor city planning and the impact of modern-day human vehicles of pollution, Cairo is unfortunately a place where one quickly becomes ill.

Moderate climates   As found in Milan these have monthly mean temperatures in the coldest months of −15°C and in the warmest months of up to 25°C. Temperatures around 20°C are rarely accompanied by humidities over 80%. However, there are times when the influence of heat from gases being expelled by industry and vehicles can change the natural climate. These are factors that must also be taken into account.

Cold climates   These tend to have high relative humidities in winter with minimum temperatures of −40°C and even colder. It is the reason why the Swedes have developed houses with very high thermal insulation values. Often when they come to export these houses, which are so right for their own climate, they forget to ask important questions about the climate in other lands.

We shall refer to climatic analysis in our outside−inside design example in Chapter 6. One shortcoming of most climatic analysis methods is that the climate of the district only is described

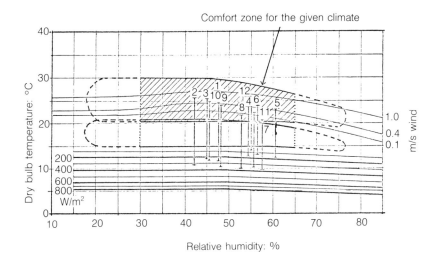

Fig. 4.10 Bioclimatic chart.

Healthy Building Code Proposal 1
The building must take nature into account. Where possible the existing vegetation should remain and new trees, shrubs, grassland and water should be replaced in a natural harmonic way.

In New York, the fresh air inlet some 40 floors up on one skyscraper block was found to be opposite the exhaust of a newly built adjoining block. The air-conditioning system of the original building became overloaded and collapsed.

(meteorological observations are deliberately kept free of local disturbances) and we do not obtain a true indication of the micro-climate of the building site.

The term micro-climate in other areas of applied science is used to describe a zone of the order of magnitude of 10 mm (e.g. a plant leaf). With a building site we have to think of an area of $10-100$ m². The micro-climate of a location is influenced by the soil permeability and soil temperature, as these affect the flora, which in turn affect the climate; by topography (elevation, slope, orientation, and exposure) and by vegetation. Plants both depend on and influence the climate of an area. This is of particular importance where parkland or woodland are intended to be removed by town planning applications, or localized building constructions.

Any planning area should be divided into climatically distinct zones, or sub-regions, taking into account shape, lie of the land, riverways, old industrial sites and geological disturbances.

Man-made environments can create adverse meso- and micro-climates of their own, modifying the macro-climate to a degree depending on the extent of man's intervention. In cities the meso- and micro-climates are a function of many interacting activities.

Anyone who has walked down Manhattan in the month of March soon learns to dive into bars, shops, or around the nearest corner as wind hits the building surface at speed and is then deflected down on to the pavement. The now demolished Bull Ring in Birmingham, England was designed as a subterranean shopping centre. Soon the shops went out of business due to the swirling currents of wind that had been directed to the spot by the arrangement of the building shapes.

Another example is to check on the glare that can be received from reflection from buildings with glass facades. In the Dutch city of Utrecht a massive structure of reflective glass caused the owners to purchase adjacent property and clothe it in a similar garb to reduce the high cost of legal actions that stemmed from environmental annoyance.

On a small but sometimes dangerous scale are the bit-planning methods of the small contractor who engages a local heating firm to install a gas boiler that discharges the products of combustion without due regard to the positioning of nearby ventilation 'fresh-air' ducts. This example is so widespread as to be seen as common practice.

Traffic noise may be present, together with the discharge at street level of fumes, coupled often to high exposures of solar radiation from bitumen-paved areas which can reach a

Fig. 4.11   Car exhaust fumes (photograph by Simon Holdsworth).

Healthy Building Code Proposal 2
Planning legislation must include provision for constant environmental monitoring of all discharges, and for such discharges to be reduced at source.

temperature more than 40°C higher than air temperature, which in turn can increase the volatility of the discharges to levels that can kill people.

Composition of air

 1 Oxygen
 2 Carbon dioxide
 3 Moisture
 4 Combustion gases
 5 Other noxious gases or fumes
 6 Irritating gases
 7 Radioactive gases
 8 Dust particles/other dust
 9 Ions
10 Biological material
11 Static electrical fields

Taken individually these items have their own influences, some benign, others less so. It is when they are combined that the resulting mixture is infinitely more dangerous.

Apart from conducting external noise surveys, designers should also institute monitoring of outdoor air quality, to build up information on the chemical, biological and other elements at points of concentration around a building development. Such information is becoming an essential factor in any build-up of environmental design criteria, and could well be undertaken by a new branch of Environmental Monitoring Agents. As well as tests for external air quality, the monitoring must also include a critical assessment of vehicular traffic, electric railways, industrial and other polluting agencies (e.g. hospitals, graveyards, toxic dumps and rubbish deposits), all of which are the breeding grounds of contaminating elements that we allow to be dispelled into the air in the belief that they will quickly become diluted and lost into space.

Design note: gas and dust pollution

	Traffic	Industry
Sheltered positioning of all ventilation intake openings (high up if possible)	X	X
Reduction of air input during traffic rush hours	X	
Reduce air input on days of heavy pollution or unfavourable wind direction		X
Protective belt of trees/bushes/climbing plants on walls; turf roofs on houses and small buildings	X	X

What's the damage?
The relative cost of sending freight by lorry and rail (units per tonne/km)

	Lorry	Rail
CO_2	0.22 kg	0.05 kg
NO_2	3.60 g	0.22 g
CO	1.58 g	0.07 g
Hydrocarbons	0.81 g	0.05 g
Soot	0.27 g	0.03 g
SO_2	0.23 g	0.33 g

The Observer, UK, April 1990

Controlling the input of fresh air can be acceptable where small quantities are involved. Higher ventilation rates are now being encouraged for improved indoor air quality, because of internal contaminants (see Section 4.2).

New design ideas will become necessary so that we can cope with the increasing man-made external pollutants. For example, in city areas where there is a high concentration of traffic fumes (e.g. as found in London or Tokyo) it may be feasible to install inflatable reservoir storage bags containing air collected at night from higher levels of the building. Such bags are controlled by external weather/pollutant sensors, for use in cut-back periods

Fig. 4.12 Ruhr industrial area, Duisburg, Germany (photograph by L. Delderfield).

Smoke and SO₂. From 1951 to 1975 on 1200 sites in the UK smoke had been reduced from 2.36 to 0.6 million tonnes, due to the Clean Air Act and the use of natural gas. However, sulphur dioxide had risen by some 0.34 million tonnes and was continuing to rise.

In 1971 the UK Government announced that SO₂ was 'unlikely to cause much damage to health'. This perceived view was proved wrong.

on the main fresh air inlet to the building's ventilation system, to reduce the high concentration of polluted fresh air.

Another possible solution is for the fresh air to buildings to be introduced through earth labyrinths. The idea is not new. In 1876 a John Wilkenson patented a 'sub-earth ventilation system' to cool a dairy. Some ten years later, underground tunnels were used to ventilate and cool soldiers' barracks in India. Cool fresh air was used to obviate the need for full air-conditioning at the Royal Academy of Music when I brought the air through a labyrinth. The interesting development of ground labyrinths comes from Germany and Poland (see Chapter 5), where scientific tests have found that bacterial impurities in air passing through such ground exchangers were reduced by half.

Composition of soil

The main constituents are:

1 Soil temperature
2 Ground water
3 Noxious components
4 Biological material

An accurate investigation of the structure of the soil and the composition of the land is essential if any building that is to be erected on such land is to be healthy.

Soil sampling and soil mechanics are well-established sciences for the determination of structure and composition of land and how it affects foundations, tunnelling and the introduction of sewers and conduits.

The position of ground-water can alter throughout the year. Ignorance or lack of sufficient information can lead to flooding of deep trenches. An unforgettable experience was on a housing site in Corby, Northants, that led to the undertaking of micro-climatic studies and the establishment of ECHOES.

In 1972 advice was given for a district heating scheme to have the service grid of high-pressure hot water pipes set above the main ground level in a special service duct that would, once covered with earth, form noise barriers, wind screens and areas of privacy for the tenants. The advice was based on local knowledge that in summer-time any water was several metres below ground. In winter, however, when the air temperatures were 10−15°C lower than for other areas of Midlands England,

and tundra conditions prevailed, the ground-water was only a few millimetres below the surface.

The advice to build the district heating pipework system above ground was ignored. A pipe-in-pipe below-ground system was declared to be cheaper, requiring many pumping pits and that the pipes should be installed during the summer period. Again, the advice was ignored by both client and main contractor. The result was multiple mains fractures due to the ingress of water, and eventual removal of the district heating system. This created a social cost to the community as well as giving district heating a bad name. It could easily have been avoided if both client and contractor had understood the need to think in whole terms, and to listen and heed the advice of the building services consultant.

Radiation

An unpublished UK Government report has recommended that local councils should be required by law to compile lists of potentially contaminated land. Unfortunately, the report also calls for reversing the caveat cmptor, *'let the buyer beware', principle.*
Construction News, *UK, June 1990*

Ground-water courses and areas of geopathic stress are referred to in this section. Subsidence and rock movements are important factors for civil engineers to deal with; however, in respect of healthy buildings, the presence of noxious components and biological material is of more immediate concern.

Radon, a noble gas which is formed when the radioactive substance radium disintegrates, is a major noxious component found in the ground and in ground-water, tap water and building materials. While radioactive radon is not referred to in the current UK Building Regulations (fire precautions take up 88 pages), and while it alarms people because it is dangerous, we await the results of a major study to find out whether it is even more of a risk to human life than fire.

While certain investigatory studies are under way in the USA and the UK, it is in Sweden that the largest radon study so far is taking place. The incidence of lung cancer in people living in a radon-contaminated house is being plotted with the help of 4500 test subjects in 109 municipalities.

Previously the common denominator for international investigation has concentrated almost exclusively on results from groups at extreme risk (e.g. miners). Those studies which have been conducted into the relationship between lung cancer and radon in dwellings have been pilot studies with a view to testing new research methodology.

On average the indoor radiation dose from radon daughter products in Sweden is ten times higher than the fall-out from Chernobyl.
Swedish Council for Building Research

It is not radon itself that poses the greatest threat to people, but the so-called radon-daughter products. Radon decays through a succession of decay products, producing metallic ions. These decay products then attach themselves to particles suspended in the air. When inhaled these decay products rest in the respiratory tract. The subsequent decay produces alpha particles which increase the health risk.

Radon wells

Radon wells are currently being used with success in Sweden, especially if installed where the building rests on a gravel outcrop. In a well 800 mm in diameter and 4 metres deep, a one-off installation reduced the radon daughter content of 15 four-bedroomed houses to as little as 2—4% of the original concentration. The total cost was the price of an expensive washing machine. The advantage of a radon well is that it takes up no space in the building, there is no irritating fan noise, and there is a potential for the device to be driven from a photo-voltaic assembly.

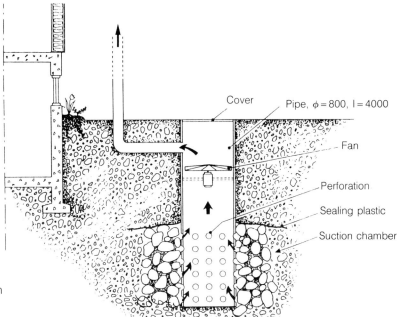

Fig. 4.13 Radon well, Sollentuna, Sweden (source: after Swedish Council for Building Research, 1984).

The designers of the radon well from the Institute of Building Research, Gavle, Sweden, emphasize the necessity of continuously recording radon daughter concentrations by the use of specialist meters and competent trained staff to handle them.[1]

Building tips

1 Pillar foundations or thick concrete foundations should be used, with loam/clay high impacted thermal insulation 'stuffing' boxes around all penetrating pipes, cables, ducts, etc.
2 Building materials with high concentrations of radioactivity should be avoided.
3 Long-stay living space below ground should be avoided.
4 Concrete underslabs should be laid in one piece, and be 80–100 mm thick with a stiffened edge. The bedding must be stable to avoid unequal settlement that could cause cracking. A thicker, reinforced slab is preferred.
5 Radon-resistant and radon-proof constructions can be obtained with a permeable bed of coarse gravel below the house and ventilated through drainpipes. Although such drainpipes can end in the footings above ground level, allowing natural ventilation, it is better to make a vertical conduit from the drainpipes to rise up through the house for discharge above roof level.
6 In basement construction (where the lower section of the walls is sealed with a membrane) blockwork with holes on the outside of the material should be used. Drainpipes are also used to exhaust radon gases.
7 Crawl spaces below a building are also another efficient way to provide a radon-resistant dwelling, if the bottom joists and the crawl space are well insulated and the soil is covered with a plastic net.
8 Where exceptionally high concentrations occur (+ 70 Bq/m^3) 5 mm diameter drainpipes should be laid below the net with a connection to a remote fan sucking the high concentrations away from the base.

All the recommendations given should be discussed with the

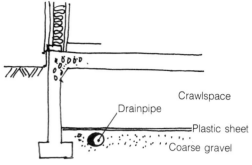

Fig. 4.14 Ventilation for a crawling space. A perforated drainpipe is placed in a layer of coarse gravel below a plastic sheet with a possible connection to a fan.

Healthy Building Code Proposal 3
National governments should, as part of their planning codes of practice and building regulations, monitor, record and regularly update pollutants affecting the outdoor environment on a regional, sub-regional and district basis.

local regulatory authorities, and continually monitored by these same authorities, or in some countries with the addition of an independent check. A radon diagnostic list is given in Chapter 6.

Toxic and biological waste

Biological materials, dangerous chemical wastes (including military) heavy metals and other dangerous substances are found in soils and buried deep into the earth. Detailed investigations of such substances are also critical for the healthy positioning of any building, no matter what its use, and for the determination of measures necessary for removal of the noxious contaminants, for protection of those undertaking the removal, and to find ways and means of disposal.

Estimating the extent and complexity of soil pollution is based at present on knowledge regarding the propensity of substances spreading out into the flow of ground-water. By 1992 the Swedish Geotechnical Institute, Linkoping, Sweden, will have an advanced method of estimation established that will enable calculation to be made on concentrated profiles in different soil strata.

Certain species of reeds have been developed as a cheap and effective method to clean industrial waste land of toxic chemicals. Developed since 1970, root zone biotechnology has been invented by Professor Kickuth of Kassel University, Germany.

Instead of removing toxic earth to be buried in someone else's backyard, a strain of reeds known as phragmites are grown. These reeds absorb oxygen through their pores above ground in air and transport it to the roots zone where it enters the surrounding soil and provides a hydraulic pathway. This simple system, which looks like a stretch of marshland, extracts nitrates, phosphates, phenols, hydrocarbons, mineral waste and harmful bacteria. The system is working well in a number of European countries, with some systems under way in the UK.[2]

Sick land can be managed back to health, in the same way as sick buildings can become healthy. Over the past 20 years a significant amount of research has been accomplished that shows that forests and arable land can be used in land treatment of waste water for direct recharge to the ground-water table. Although waste water contains a lot of nutrients that can be used, it also could contain trace metals and toxic substances, a matter for public concern. Such concern has related to agricultural systems because the crops may go into animal feed, and trace metals thereby enter the human and animal food chain. However, research undertaken by Professor Dr William Sopper, environmental scientist and Professor of Forest Hydrology at Penn State University, Pennsylvania, USA, has shown that waste water can be applied in forested areas with very little health risk to animals and people, and that the areas can be used as recreational green belts with added amenity value to property.

In 1962, the Penn State University, which owned a waste treatment plant, found that due to rapid population growth in the university and the nearby town of State College, the effluent being dumped into the only surface spring in the area, gradually turned a beautiful spring into an open sewer. During this same period the area was experiencing a seven-year drought with a deficiency of one-and-a-half years natural precipitation. The only water supply in the area was ground-water and this was fast being depleted from wells installed by both town and university. While this was happening, millions of gallons of sewage effluent were being discharged by the local treatment plant for rapid transportation to the Atlantic Ocean some 200 miles away.

Dr William Sopper[3] and his colleagues, in an endeavour to reverse this piece of unsustainable nonsense, evolved the concept known as a 'living filter'. Waste water was made to work with the natural system to irrigate forests and to grow trees, or to irrigate crops and then have the waste water renovated by the natural system through degradation by micro-organisms, chemical precipitation, ion exchange, biological transformation and the uptake of nutrients by absorption through the roots of growing vegetation.

The combined population of the town and university is 70,000 people, and the university treatment plant of trickle filter design has a capacity of 4 million gallons a day. After chlorination the water flows by gravity to a pumping station where the waste water is moved to two sites. The first site is primarily agricultural, and the waste water is usually applied in the summer months. The crops grown include reed canary grass, which loves a lot of moisture and is harvested as an animal feed. Corn is grown for silage and also to take up the maximum amount of nitrate, since the main worry of any land treatment scheme is to prevent nitrates entering the ground-water system.

Agricultural systems were found easier to manage because applying waste water for irrigation all summer applies, in the case presented, 200 lb of nitrogen. The same amount is taken off at the harvest, leaving a very small amount to leach into the ground.

Forest systems were found to be more tricky to monitor and to manage, particularly for nitrate levels. Forest eco-systems have different capabilities for accepting and renovating waste water; also, trees take up very little nitrogen relative to their size. However, it was found in this climatic region that the white spruce species of tree developed diameters twice the size of those not irrigated, and that a mixed eco-system of white spruce and herbaceous vegetation renovated the waste water without any harvesting. The trees could provide a crop mainly for pulp.

Throughout the whole 20-year period of irrigation with waste water the quality of water was carefully monitored for trace metals. Table 4.2 shows the quality of waste water for seven major trace elements.

The figures are all very low because most of the heavy trace metals end up in the sludge produced at the treatment plant. Most trace metals are not very soluble in an alkaline solution

Table 4.2 Typical chemical composition of the waste water (Dr W. Sopper)

Constituent	Average concentration (mg/l)	Average annual application* (kg/ha)	Total amount applied† (kg/ha)
Cu	0.068	1.1	22
Zn	0.197	3.2	64
Cr	0.022	0.4	8
Pb	0.140	2.3	46
Co	0.040	0.6	12
Cd	0.003	0.05	1
Ni	0.050	0.08	16
pH	7.5	—	—

*Total amount in forest eco-systems which received waste water applications at 5 cm/week (2 in./week).
†Applied over 20-year period (1963–1982).

and most sewage effluent is neutral or alkaline. The annual amounts of trace metals that were applied on the land due to irrigation of 2 inches per week for 30 weeks per year were 1 lb of copper, 3 lb of zinc, and 0.5 lb of cadmium, the element that presents the greatest health hazard, accumulating as it does in the kidney, liver and bone. It is why fish pounds in such areas are attractive, since we discard the organs of fish and hardly eat the bones.

The advantages are the opportunity for water pollution abatement, with the ability to recycle and the beneficial use of the nutrients contained in waste water. It also allows the replenishment of local ground-water supplies and the preservation of open space and inner city parkland. The irrigation system at Penn University is 70 m away from the hedgerow of a housing development where homes sold for $US70,000 in 1963, and for $US190,000 in 1990. There is now a resort area, conference centre and golf course, all irrigated with waste water.[3]

It can well be seen that factors of influence, if they become malign, can be controlled in the same way as we have developed methods of control for sun shading, wind deflection and other natural external factors of influence.

Radiation

1 Very low-frequency radiation
2 Radio waves
3 Heat radiation (solar)
4 Visible light
5 Ultraviolet light
6 Gamma radiation
7 Cosmic radiation

Healthy Building Code Proposal 4
Recommendations for buildings close to EMF sources should include:
- sufficient distance to be allowed between buildings and EMF source
- provision for earth beaming
- sleeping areas to be over 1 metre from any local low source

Since the 1930s there has been an enormous use of mains electricity and electrical apparatus, followed by an explosive spread of broadcasting: first radio, then television, radar and communications networks. The effects on the health of people and animals are only just beginning to be realized. Geopathic zones (areas where subterranean watercourses cross over, and geological earth faults occur) exhibit magnetic charges as well as localized increases in microwave and infra-red emissions, a concentration of gamma rays and slow-moving neutrons as well as changes in VHF field strengths. Sources of manufactured fields include generating plant, high-tension power lines, domestic wiring and apparatus used in the home and office (see section 4.2).

Everyone on earth is continuously surrounded by 50 Hz fields (60 Hz in the USA), and the density of radio waves is now 100−200 million times that reaching us naturally from the sun.[4]

In northern Minnesota in the early 1970s farmers became concerned at the reaction of gigantic high-voltage transmission lines with 745,000 V of electricity being sent through them. When one farmer took a fluorescent light tube and stood a few hundred metres away from the transmission line and it lit up he knew that something was 'unnatural'. This particular farmer started to shoot out the insulators on the electric masts, which resulted in a small environmental war between farmers and the electric utility companies, who flew helicopters, Vietnam style, during the replacement of the lines. We have all taken electricity as a useful and benign gift. The ability just to be able to 'flick the

Farmers have for generations known that there are certain places in a barn where a cow becomes ill, is restless, and cannot give much milk.

switch' resulted in the dumping of solar water-heating panels that had been used in the American western States 100 years ago. Governments have constantly demanded the production of electricity no matter what damage the production of this powerful force has had on the ecological balance of planet earth. Electromagnetic field forces are powerful and can damage our health. The designers of distribution systems must begin to seek ways of reducing that damage, perhaps by putting the transmission cables within special ducts within the earth. We shall still need transportation of electrical power even when we generate electricity by wind and wave and solar power.

Electromagnetic rays have a wavelength of the order of 1 cm and can be measured by purpose-made instruments. Such instruments can identify both electrical and magnetic field components. The biological effect of these fields can be demonstrated using an earthed voltmeter to measure the electrical potential generated in the subject by exposure.

At the spot of the ray there is a charge of the ion environment and of the earth's magnetic field, whence a much stronger radioactive ray can be observed. (The reception of FM stations can be locally stronger or weaker at spots where 'earth rays' are located.)

There are two types of electromagnetic ray, natural and artificial. Natural rays occur in underground watercourses, geological faults, mineral veins, oil and gas reserves, geomatic zones, and growing zones. They are also part of the human DNA system and are important not only for our own communications network, but also because they enable us to heal ourselves.

Artificial rays are now occurring from high-tension cables, tram and train cables, radio stations, television and radar, resonators, all electrical equipment, and, of importance for human beings, the mind (meditation, music, sound, the DNA structure).

Curry and Hartmann nets are means of plotting the points of positive and negative ions on a map. Natural geopathic stress areas have for long been determined by dowsing, an ancient art that was practised first in China many thousands of years before Christ, enabling earlier man to live comfortably with natural radioactive force fields. The interaction of artificial, man-made electromagnetic radioactive forces is the concern of this book.[5]

There are few urban areas in the world today where investigation would show that 70−90% of all electromagnetic forces were not artificial. According to the German physicist Reinhard Schneider, everything in our world has a polarity: man−woman, Yin−Yang, positive−negative. In the new science of radio-aesthetics, left polarity is unfavourable, whereas right polarity is in certain circumstances favourable to man. The intensity of a ray is the graduator of its biological activities. A high-intensity 3,2,1 ray means a strong influence on the human being. Conversely, a 5,6,7 ray has reduced influence.

The influence on all living beings can be different. It depends on the type of radiation, force field (polarity) and intensity. Natural earth rays can be immensely powerful. The interaction of watercourses creates a negative discharge field (Yin), whereas a geological fault creats a positive discharge field (Yang).

Yin and Yang are the opposite poles of intensity. According to a number of German and Dutch architects who have made detailed studies of these new forces, height above ground appears to make little or no difference to the power of these changes in intensity. Thus a person living on the top of a tower block of flats is likely to be affected as much as a person living on the

Fig. 4.15   Electricity pylons (photograph by L. Delderfield).

Healthy Building Code Proposal 5
Establish a radon diagnostic list similar to that of the US Environmental Protection Agency, 1987.

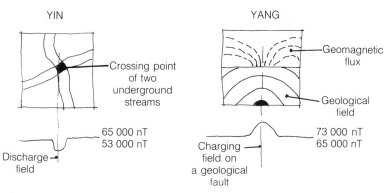

YIN

YANG

Crossing point
of two
underground
streams

Geomagnetic
flux

Geological
field

65 000 nT
53 000 nT

73 000 nT
65 000 nT

Discharge
field

Charging
field on
a geological
fault

Fig. 4.16 'Geomagnetic flux' known as a geopathic disturbance. The UKCIBSE Guide and the US ASHRAE Guide currently have no recommendations for electromagnetic radiation.

'Geomagnetic flux' known as a geopathic disturbance

ground floor. (In fact, there is now serious evidence to suggest that the radon elements in concrete can aggravate a radioactive/ geopathic illness.)

Geopathic zones are characterized by variations in terrestrial magnetism, for the earth's field is not uniform but exhibits many highly localized distortions, some random, some fairly regular. These occur over geological faults, caves and underground water courses, places where the earth's natural and beneficial field increases or decreases rapidly (a high magnetic gradient). Flows of water underground produce the largest effects. Sewers and drains can be as great a hazard as underground streams.

The Amsterdam architectural practice, Bio-Logisch Architekten Kollektief,[6] have undertaken research and found that:

1 Geological fault points can also cause structural cracks in buildings.
2 Some electromagnetic forces cause good sleeping conditions, others bad sleeping conditions, and bed positions must be altered (see Section 4.2).
3 High-intensity rays (3,2,1) can, in combination with other force fields, cause chronic illness, and cancer. These points on the Hartmann/Curry grid are known as cancer crossings.[6]

Healthy Building Code Proposal 6
Establish a cancer register to make fact-finding studies more viable

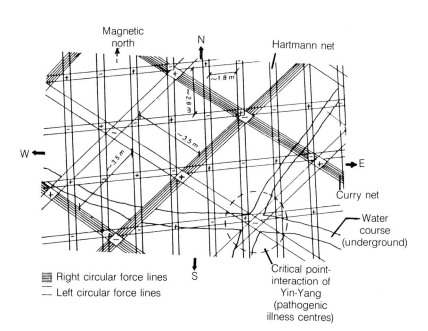

Magnetic
north

N

Hartmann net

~1.8 m

~2.8 m

~3.5 m

W

~3.5 m

E

Curry net

Water
course
(underground)

▦ Right circular force lines
— Left circular force lines

S

Critical point-
interaction of
Yin-Yang
(pathogenic
illness centres)

Fig. 4.17 Hartmann–Curry net.

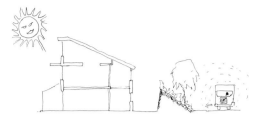

Fig. 4.18   Earth beaming, reduces traffic noise; trees and shrubs can dilute fumes.

Noise and vibration

Of all our senses of perception, the ear has a greater field for gathering information on what is happening around us than the eye or the nose. We are born in the shape of an ear. While we are still within the protected fluid of the womb the ear is the first of our organs to be fully operational. We have already started to hear the sounds in the world outside. A child soon recognizes the voices of his parents. Sound is part of our well-being. Noise contaminates natural sound melodies. Ultra-sound frequencies can be both positive and negative.

Noise is a contra-sound: our organs feel it and react. With the concentration of populations into cities, traffic noise has become a major source of environmental discomfort. Mixed noise sources of roads, aircraft and trains create areas which are unacceptable for building, both in human and economic terms. High-frequency sound levels are emitted from diesel lorries, even when they are stationary; another source is the vehicles of people who drive without consideration for others. A single driver who suddenly accelerates his car to 4000 rev/min makes as much noise as 32 other drivers at speeds of 2000 rev/min.

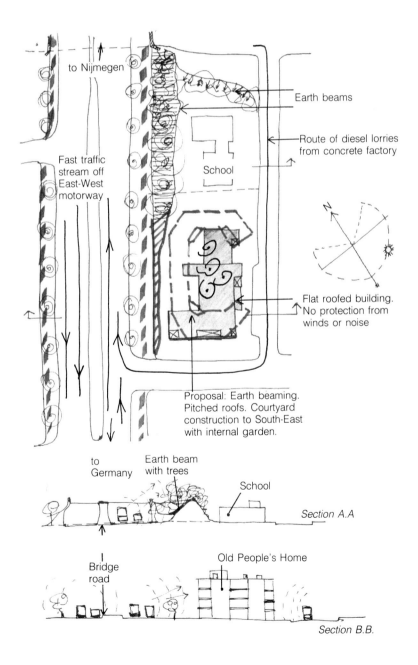

Fig. 4.19   Old people's home, Nijmegen, The Netherlands.

Healthy Building Code Proposal 7

Noise reduction measures:

- earth beams
- noise barriers
- design shape and form
- protective belts of thick trees (coniferous) and thick bushes
- heavy and porous building materials such as loam
- turf roofs
- sensitive planning

Physiological effects, starting at 65 dBA with mental and bodily fatigue, are well established. This is a typical city noise level. Main-road kerbside noise at 75 dBA is over twice as loud, and motorways nearly double again. Street noise can reach 90 dBA, causing heart stress.

As soon as the decibel volume exceeds 80 dB the blood pressure starts to rise; the stomach and the intestine operate more slowly, the pupils of the eye become larger and the skin becomes paler. Children in particular can suffer from high blood pressure due to high noise inputs, with resultant difficulty in coping with arithmetic and solving simple problems.[7]

A main feeder road to the European motorway system runs close to the apartment block where these words were written. The weekday noise level is rarely below 70 dBA. Close to the road the city council has allowed a Humanist Association to fast-build a small apartment block for old people. There are no earth beams or other silencing agents to protect it. Double glazing has been installed in a building constructed of concrete with flat roofs. There are no overhangs. The site is close to a primary school, and could, with earth beams and trees, give the school added quietness. A more tranquil site would have been more suitable for the old people's flats. Instead we have a classic example of erecting an unhealthy building.

In 1973 my building services consultancy was commissioned to design the total services (including acoustics) for staff housing for British Airways at Stanwell, Middlesex, England. The building site was in the direct flight path of aircraft taking off from London Airport. The sound levels were in the +90 dBA range. Using the ECHOES system of design, the architects took on board the resultant proposal for a set of buildings built on the Chinese courtyard principle: thick external walls, small triple-glazed windows, roof overhang, and a sheltered internal courtyard with gardens and trees. The internal rooms could be simply ventilated by natural means. The aircraft noise was muted to acceptable human limits.

Local planners dumped our 'healthy' ideas on the spurious basis of space ratios and commitment to Ministry planning guides currently in vogue. The result was an internal open stairwell, with internal mechanically ventilated kitchens and bathrooms, and all the living rooms on the outside. The cost of acoustic protection (covered by a Government grant) increased, whereas this same money or less could have produced a better solution in the first place. The inhabitants were stand-by pilots and air hostesses requiring, when they were home, some peace from strain, stress, noise and vibration. None the less the buildings

Fig. 4.20 Low-cost staff housing for British Airways, Stanwell, Middlesex. The architects, Edwards-Stephan Associates, London, preferred the ECHOES solution (photograph by R. Bryant).

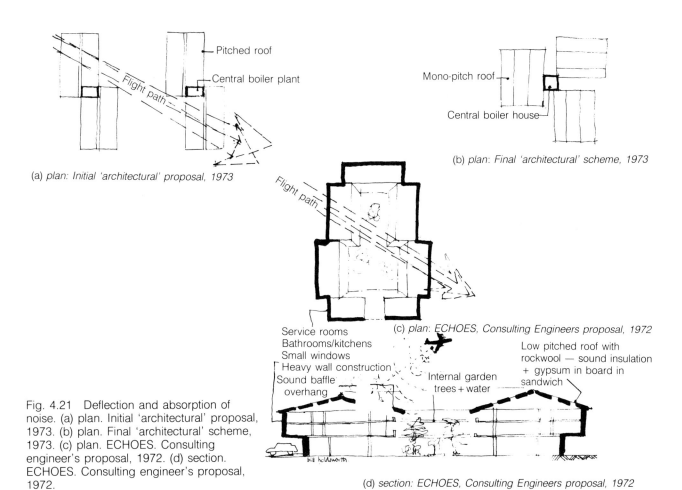

(a) *plan: Initial 'architectural' proposal, 1973*

Pitched roof

Central boiler plant

Flight path

Mono-pitch roof

Central boiler house

(b) *plan: Final 'architectural' scheme, 1973*

(c) *plan: ECHOES, Consulting Engineers proposal, 1972*

Flight path

Service rooms
Bathrooms/kitchens
Small windows
Heavy wall construction
Sound baffle
overhang

Internal garden
trees + water

Low pitched roof with
rockwool — sound insulation
+ gypsum in board in
sandwich

bill holdsworth

Fig. 4.21　Deflection and absorption of noise. (a) plan. Initial 'architectural' proposal, 1973. (b) plan. Final 'architectural' scheme, 1973. (c) plan. ECHOES. Consulting engineer's proposal, 1972. (d) section. ECHOES. Consulting engineer's proposal, 1972.

(d) *section: ECHOES, Consulting Engineers proposal, 1972*

won an architectural design award, which illustrated how little aware were award committees of problems of environmental pollution.

A number of methods are used to deflect and absorb noise. One is the way in which we construct our buildings, the other is the use of barriers.

Buildings can be constructed of heavy and porous materials. Concrete is used as a mass material to deflect and stop noise penetration. Buildings can be shaped as in the NMB Bank Building, Amsterdam (see Case study 7, Chapter 5). Here the deflective shaping not only resolved acoustic requirements, but also benefited the daylighting needs of the inhabitants, as well as creating a pleasing structure for the eye. The use of other materials such as clay loam, impacted between lightweight walls, and turf roofs on houses, and constructing small buildings are beneficial ways of absorbing noise and at the same time protecting the inhabitants from other pollutants such as radon penetration and electromagnetic field forces. Chapter 5 shows how the use of the right material carefully designed, the designer having first looked at all the factors of influence, can result in solving more than one of the problems of environmental pollution.

Noise barriers, earth beams and protective belts of trees are answers being proposed and erected in many cities and towns. The sight of many concrete utilitarian slab blocks by heavy traffic roadways through our congested cities leaves much to be desired. There seems to be a lack of imagination on the part of planners and contractors. Concrete need not look like a battlement against tanks. Such areas of slab soon become the walls for the city graffiti artists, the only place that I can accept them with some grace;

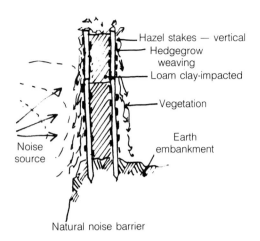

Fig. 4.22   Natural noise barrier.

however, I believe that with the integration of artists, sculptors and landscape gardeners, such canyons, that unfortunately become necessary while we drive new transportation ways through heavily populated areas such as east London, can become a haven for trees, vegetation and birds, and places of rest on the quiet side of the barrier. A recent intelligent departure can be seen on the European east—west motor route close to Nijmegen in The Netherlands. Using a technique that could once be found for the protection of an Iron-Age encampment, an embankment was formed of two lines of hazel stakes with other horizontal hazel branches woven to form a hedgerow. The section in between was filled with impacted clay loam. Within a few short summer months the whole embankment was overgrown with wild flowers and grasses. The noise reduction quality is as high as if concrete had been used. The erection time was less, the energy input both in time and material terms was less, and the whole result is more in keeping with the nature around us. Naturally one may have to think differently in a central city environment, but it is a start to look at such problems from an alternative point of view.

Protective belts of trees forming thick forest areas not only provide a sustainable product, but also can be irrigated by waste water. It is remarkable how a country with such a high concentration of people as The Netherlands can, by the provision of forest belts, create noise protection for its inhabitants. Many of the multitude of small towns and villages cannot be seen from the main routeways. Neither do you hear the main traffic noise when you find yourself in the places of habitation. Closer to home, the use of earth beams gives both protection from winds and noise, and can be used for the routeways of city infrastructures such as district heating and cooling systems, places where transformers can be buried, cut-outs for bottle banks, and other rubbish collection areas. Here are places to sit out of the wind, and, with the provision of trees and shrubs, a place of birds and wildlife.

Noise need not penetrate our living areas. Provision of such areas of protection is important when designing for healthy buildings. The area of design responsibility does not stop at the building boundary line.

Although it is outside our immediate remit, the reduction of noise and vibration at source is the first answer, and if such legislation were enacted it could save a great deal of other social and economic expense for national budget sheets.

In 1941 when the world was busy with using its precious supplies of raw materials of oil, coal and steel for the purposes of war Henry Ford unveiled his biological car. The body of the car was made of soybeans. The tank and the fuel came from corn, and the wheels and steering mechanism from the flower goldenrod. This organic car was lighter — and dents could be knocked out; it was cheaper to run, warmer in winter and cooler in summer. It was also very quiet. The flood of cheap oil on the world market after the Second Wold War destroyed Henry Ford's dream. Perhaps it is time that we dreamed again . . .

Relationship to other buildings, surroundings, shade and openness

This sub-section of external factors of influence is referred to in other parts of this book, since such elements are of importance for the particular building we may ourselves be dealing with. What is important to the client and developer as well as to the planning team for the many gigantic developments that seem to take place in city areas throughout the world (e.g. London's Docklands and the area of north London close to King's Cross mainline railway station) is that the influence of other buildings and more important, the uses of other urban space is critical for the creation of an individual healthy building. We have also to begin to think in terms of healthy cities at the same time.

Vegetation

Natural vegetation should wherever possible be reintroduced on to any building site. The Dutch Building Code of Practice has since 1910 made it a legislative condition that wherever possible landscaping with trees, shrubs, flowers and grass, as well as water (ponds, small lakes, streams) are introduced into all urban and rural building developments.

It has been a tradition in The Netherlands since the 1930s that most housing, office and industrial sites co-operate to make compost from household waste through a common system of compost collector plants. A recent problem of toxication of such compost is now in the process of being overcome by the introduction of new legislation to control the amount of toxins.

For a country that has such a high population concentration for available landscape, The Netherlands is a pleasure when it comes to the imagination and dexterity of planning the external micro- and meso-climates of cities. Transportation routes are segregated. Towns and villages are protected and hidden from the main routeways by thick bands of sustainable forest land. Within these woods, some of which are irrigated by waste water, there are public gardens where spices for cooking can be picked. In the townships themselves trees and plants are planted first since they take a longer time to grow, before the introduction of buildings. Of course such ideas are not always fully realized, and at times, builders try to cut the boundary edge, and lack of inspection reduces the amount of landscaping.

The role of plants in developing healthy atmostpheres has long been known, but strangely confined to rural and garden-cities, a unique breed of smaller towns outside the main metropolitan areas. Putting trees and plants firmly back on the menu of design is as essential as any other proposal contained in this book to counter the cocktail effect of the many pollutants that have been listed that now contaminate our air, water and soil.

The overlapping of industrial and residential urban zones often creates uncongenial and unhealthy surroundings. The term 'green field site' is a misnomer, often only referring to a site area free of any other building, or situated on the outskirts of a fast growing city. The construction of healthy buildings requires trees and greenery as an intrinsic factor of its architecture;

Fig. 4.23 (a) Different shapes of leaves suitable for dust trapped. (b) Different shapes of canopies suitable for landscaping colonies.

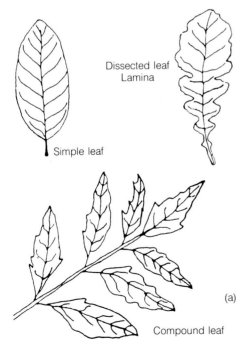

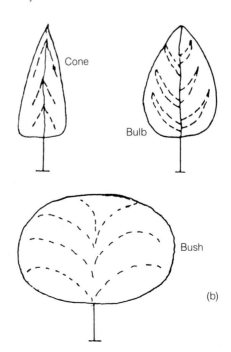

Fig. 4.24 Trees (photograph by Bill Holdsworth).

equally, a comprehensive approach is needed in selecting the right tree and shrub to filter the pollution.

Such selection is enabled by what has been called an ability index (AI)[8] for landscaping. This index takes into account the various physiological parameters of trees which include such elements as ascorbic acid, nitrogen and relative water content.

The analysis of gaseous air pollutants, SO_2, NO_2 and dust fall is known as the pollution stress rate, which is then multiplied by the population density of animals and people:

$$AI = \frac{\text{area of tree canopy } + \text{ internal physiological resistance (IR)}}{\text{pollution stress } \times \text{ population}}$$

The index obtained from deciduous trees is multiplied by 2 as these trees and plants produce a new crop of leaves annually. Plant species known to be tolerant or sensitive in one region may not be suited to another place. Climatic and regional conditions have to be taken into account. Species suited to hot, dusty climates are different from those suited to colder, wet climates.

Table 4.3 was collated for the fast-growing industrial city of Hyderabad, India.

Table 4.3   Analysis of trees and plants for the industrial city of Hyderabad

Name of tree	Canopy (ft^2)	IR	Dust deposits (g/m^2)	AI
Albizzia lebbeck	12.56	167	34.6	1.5
Annona squamosa	14.44	268	33.4	2.4
Azadirachta indica	21.98	284	37.5	2.6
Caesalpinia pulcherima	18.84	183	48.3	1.7
Cassia fistula	18.84	270	48.0	2.4
Dalberigia sissoo	12.56	279	41.6	2.5
Eugenia jambolana	12.09	127	34.1	1.2
Phoenix sylvestra	6.28	156	23.7	1.4
Pithecolobium dulce	9.42	250	76.3	2.2
Pongamia glabra	15.70	195	25.6	1.8
Polyalthia longifolia	1.53	159	22.9	1.4

Trees such as *Azadirachta indica, Eugenia jambolana, Cassia fistula* and *Pithecolobium*, with a busy type of canopy and broad leaves, are able to trap much dust and act as sinks for pollutants. Plants are also responsible for the air or wind flow and for dispersing pollutants. Those having a bushy type of canopy with more leaves have more filtration and act as a better air purifier than those plants with conical and congested leaves. A plant's response to a pollutant also depends on its internal resistance, mainly due to chlorophyll, ascorbic acid, nitrogen and relative water content.

The higher the ability index value the better the effect on filtering out the pollutants, whereas the lower values shown are for trees/plants that monitor well the quality of the external air.

4.2 Factors of influence: internal

Note: As with external factors of influence it is not our intention to reprint information that is already well documented and in common usage. Current updates, and technical data on such factors as moisture, radon and outgassing from building materials, are targeted, as well as bringing together information on the miscellaneous equipment that we bring into our homes, offices and other workplaces. Most advanced industrial countries have Factory Inspectors. It was their job to check machinery and workplaces to ensure safety to both working operatives and the building in which the machine or industrial operation was located. At present there is hardly any control of the new breed of computerized equipment that we have accepted so easily, that has been marketed through ordinary shopping procedures. A new dimension to planning is with us.

As previously stated in the introduction to the external factors of influence, the check lists for internal factors of influence should be used in a logical order. Each element is considered both individually by the designer of a particular discipline, and then again by every other member of the design team, and also where the elements of the matrix intersect and influence each other.

It is at these 'intersecting' points that many of the problems that lead to unhealthy buildings occur. The proper questions are not asked as to the usage of a space, the number of people, the determination of the actual occupier's views. Such questions are often left to the last, when instead they should be paramount to the whole design operation because the answers to such questions affect all the design criteria. We shall be illustrating our ECHOES method with storyboard illustrations in Chapter 6.

As referred to previously, one of the oldest of building illnesses with its attendant effects on the inhabitants is moisture, rising damp and humidity.

Moisture: rising damp and humidity_____

Essential items of building physics and its attendant technology are:

1 Try to keep the building dry during construction.
2 Drain off water where it appears (e.g. in areas such as the foundations, window frames and outer walls, kitchens, bathrooms and places where steam is produced).
3 Ventilate building areas exposed to humidity.

C.H. Sanders of the Building Research Establishment (Scotland)[1] found, from an extensive study of surface condensation and mould growth that affects a large proportion of the UK housing stock, the following remedies:

1 Extractor fans in bathrooms and kitchens are likely to be much more effective when controlled by humidistats than when controlled by tenants.
2 Improved insulation reduces the risk of mould in heated rooms. Conditions in unheated rooms in flats are hardly improved with insulation without heating, unlike conditions in two-storey housing where insulation may suffice.
3 Provision of insulation and whole-house heating eliminates the risk of mould growth. The cost of insulation is offset by the reduced cost of the heating system.
4 The performance of dehumidifiers depends on the room air conditions. They are effective in well-heated dwellings with problems of high moisture generation or low ventilation but are less effective in colder properties. They require the active co-operation of house-owners.

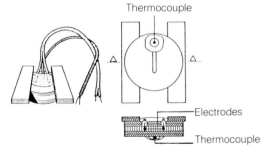

Fig. 4.25 Roundel of plywood for moisture measurement (after T. Bunch-Nielsen, Miljoteknik a/s).

Such recommendations on how to avoid condensation will seem to some readers from the USA and Scandinavia and elsewhere to be archaic. Improved thermal insulation to much higher standards is long overdue. The use of solar passive techniques in building, and materials that can breathe, are tools to stop this rot to housing. Your particular attention is drawn to the work of the Dutch architect, Renz Pijnenborgh, referred to in Chapter 5, and the Norwegian architectural group, Cobolt Arkitekter (see p. 55f).

Design of moisture-resistant structural features

The following principles are for new buildings and for the rebuilding of existing buildings: prevention, heating, and ventilation.

Prevention

Moisture should be prevented from reaching parts of the building by:

1 Inserting impermeable layers
2 Incorporating layers intended to carry away water
3 Drainage
4 Capillary breaks
5 Temperature elevation

The use of impermeable layers must be suited to applications and conditions. Correctly used, an impermeable layer will result in a drier structure than will one without. However, a structure that incorporates an incorrectly applied layer will trap moisture that would otherwise have dried out. Any impermeable layer should prevent the undesired transportation of moisture without causing unnecessary risks.

Note: This is a point for design interaction. When dealing with construction detailing for radon protection, care must be taken that what is the right solution for one problem may not be correct for another. Design interaction must occur.

Water must be led away. This applies primarily to the roof, and water should not be allowed to remain standing for long periods. It also applies to window sills, drip plates over doors and windows and at joints between parts of buildings. Water in the ground must not be allowed to flow towards the foundation structure.

It may be necessary to incorporate a capillary-breaking layer of gravel in order to prevent ground-water being drawn up, and in order to maintain a permanently dry structure. It is good practice to prevent capillary transport between moist concrete and wood by the insertion of a slate or felt insert.

Note: the radon/water design interaction occurs here again. Reference to p. 39 indicates that not only does a layer of gravel help to reduce radon gas penetration, but that it is also a factor in reducing the field of electromagnetic forces from subterranean watercourses.

Heating

Parts of the building structure should be heated where raising the temperature of a structural element or part of a building helps to reduce the transportation of water vapour. High standard insulation is one method. But the use of solar heated walls and other concepts using renewable energy is also beneficial. The importance of high insulation at all cold bridging points is also a critical design interaction activity between the architect, the structural engineer and the building physicist.

Heating of a building structure should never be intermittent. Temperature gradients change according to external wind, rain and solar patterns. This is particularly important for basement structures that should be well ventilated.

Ventilation

Ventilation should be provided where moisture can occur. Ventilation should be employed in roofs, walls and floors. Ventilating away moisture provides additional security against damage. The air should be moved, if possible by natural wind pressure and thermal forces. The use of fans should be kept to a minimum. But, as with the need for the radon well fan, the introduction of fans powered by solar photo-voltaic cells via batteries can provide a good market for inventive manufacture.

It is equally important that the ventilation system should be designed so that damage caused by convection is avoided. The air should always move from the lower temperature to the higher temperature area. While in a ventilated underfloor space, the air should be drawn in, along or through the exterior wall, and towards the centre of the building, where it can be evacuated, thus picking up an increased amount of moisture.

The occupants of any building must fully understand the importance of the term 'ventilation'. It does not just signify draughts. The occupants must also be fully instructed in the combination of ventilation, temperature and humidity control. Wherever possible such methods should become part of the natural life of the building.

Rising damp in buildings is still a major cause of unhealthiness. Unless careful and comprehensive investigation techniques are used to correct and identify design, construction and maintenance faults, appropriate effective long-term solutions will remain impossible to achieve. Capillary action within the pore structure of wall materials, sometimes combined with associated water pressures underground, will frequently result in upward movement.

Another factor is the occurrence of inorganic hygroscopic salts which derive from ground soils. Once these salts are established on exposed wall finishes and floor surfaces, additional moisture is absorbed from the air to increase damp problems. Salts have to be dried out.

Dampness in walls and ground floors also derives from other causes, unrelated to rising damp but often in practice mistaken for it. In such circumstances damp-proof courses are not needed. Such dampness can relate to defective external plumbing, leaking water pipes, lateral rain penetration and condensation effects. Surveys should always be comprehensive.

Moisture and timber buildings_____

The use of timber-framed construction in all forms of buildings is increasing. The material is healthy. It can be produced and processed with renewable energy technologies. It is a sustainable product and carefully husbanded gives to the building industry

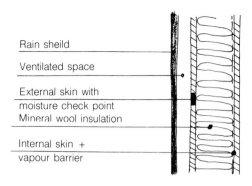

Rain sheild

Ventilated space

External skin with
moisture check point
Mineral wool insulation

Internal skin +
vapour barrier

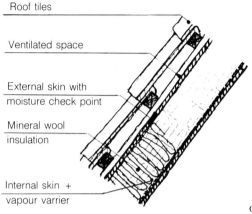

Roof tiles

Ventilated space

External skin with
moisture check point

Mineral wool
insulation

Internal skin +
vapour varrier

Fig. 4.26 Sandwich panels used in external
walls and roofs.

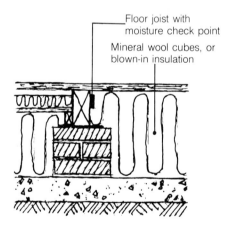

Floor joist with
moisture check point

Mineral wool cubes, or
blown-in insulation

Originally the space was insulated with 50 mm
mineral wool, or not insulated at all.

The blown-in method was developed in Denmark
in 1985; since that time the moisture content of
the floor joists has been regularly measured.

The moisture content of the wooden parts of a
crawlspace normally reaches maximum in
Denmark in August due to high humidity of the
ventilating air.

However, the moisture content stays level all the
year round when the crawlspace is fully insulated.

This check has been undertaken on many
hundreds of houses with effective results.

Fig. 4.28 Typical crawlspace under a floor.

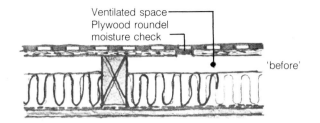

Ventilated space
Plywood roundel
moisture check

'before'

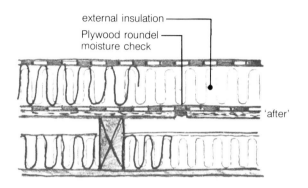

external insulation
Plywood roundel
moisture check

'after'

Fig. 4.27 Ventilated roof before and after
additional insulation.

one of its most viable products. As well as moderating humidity, timber absorbs dust particles and airborne toxins, provided it is not sealed with oil-based paints and polyurethanes.

One of the main problems with wooden structures is moisture which can lead to fungus attacks and mould growth, thereby creating unhealthy buildings and indoor climate problems. When timber is transported it requires protection. Exposed to inclement weather (wet snow is the worst) the moisture content can increase by 25%. Ordinary timber houses require a 60-day drying time, varying from detail to detail. Half-dried means half sick. If the drying time is too short, the moisture content will be high in other parts of the structure (e.g. the ground plate).

Some Swedish timber-framed construction systems have endeavoured to overcome these problems by manufacturing complete wall, floor and roof sections (including in the walls, insulation, doors and windows) in a protected factory environment, and then transporting by container to the site for rapid erection.

During the past 10 years moisture measurement components have been built into Danish lightweight timber structures for regular readings.[2] The measuring method consists of roundels made from a circular piece of plywood with two electrodes that measure the electrical resistance of the moisture. The resistance is dependent on temperature. An accuracy of $1-2$ wt% is obtained, with measurement from the same point occurring at intervals of 1 week up to several months.

The use of simple measuring devices gives building contractors and specifiers on-site recorded information which can relate to a particular micro-climate and therefore give a building developer early warning of the need for design changes and remedial action in earlier building constructions.

Compact wooden structures without ventilation have proved to function in practice, and the measured moisture content in the wooden parts is below the limits for fungus growth. Furthermore, the risk of mould growth inside the structure is very small.

Wooden sandwich panels with a ventilated climate shield can

Healthy Building Code Proposal 8
For checking moisture content a timber meter is as important to a contractor as a fever thermometer is to a doctor.

be used for rooms with a humidity of up to 8 g/m³ air at 20°C. Warm structures with external insulation are very safe from moisture accumulation, and if the thickness of the external insulation is adjusted to the humidity of the room the structure can be used even in wet buildings.

One of the big problems with all types of construction, and particularly with timber-framed buildings, is the installation of vapour-barriers: to allow a wall to breathe, and at the same time not to let the external elements through into the internal space. The architectural firm, Cobolt Arkitekter, Norway, have developed methods to reverse the normal technique of a wall constructed of external timber board, or tile and air gap: a slow-permeable masonry wall, with internal insulation and plaster or timber finish. So that as little possible interstitial condensation can occur a 'stack effect' ventilation path, achieved for instance by having an open fireplace, can draw air in through the walls. Such walls save energy because heat leaving the building warms up the incoming air.[3]

Christopher Day, in his book, *Places of the Soul*,[4] suggests that attention needs to be paid to the moisture-moderating properties of the building fabric as well as to the absence of any pollution source within it (such as formaldehyde from chipboard or plywood). Because it has to take account of ever-changing conditions of weather and occupancy this sort of construction

Fig. 4.29   The moisture transfusive house.

A   Traditional Norwegian timber-cottage
B   Pisé building in L'Isle d'Abeau, France 1985. Architects: J.M. Savignat, O. Perreau Hamburger and M. Munteau
C   Dwelling in Stavanger, Norway 1986. Architects: Cobolt. Also detail.

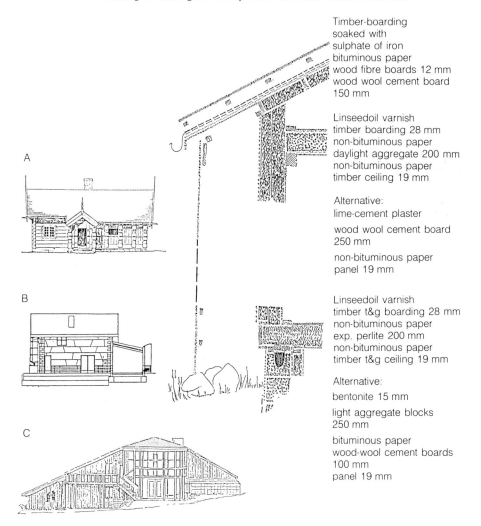

Timber-boarding
soaked with
sulphate of iron
bituminous paper
wood fibre boards 12 mm
wood wool cement board
150 mm

Linseedoil varnish
timber boarding 28 mm
non-bituminous paper
daylight aggregate 200 mm
non-bituminous paper
timber ceiling 19 mm

Alternative:
lime-cement plaster
wood wool cement board
250 mm
non-bituminous paper
panel 19 mm

Linseedoil varnish
timber t&g boarding 28 mm
non-bituminous paper
exp. perlite 200 mm
non-bituminous paper
timber t&g ceiling 19 mm

Alternative:
bentonite 15 mm
light aggregate blocks
250 mm
bituminous paper
wood-wool cement boards
100 mm
panel 19 mm

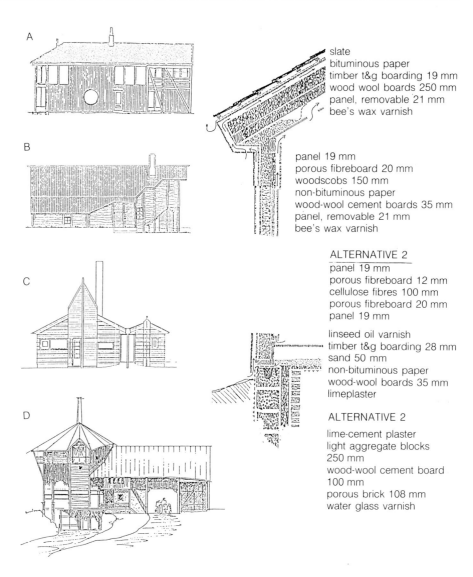

slate
bituminous paper
timber t&g boarding 19 mm
wood wool boards 250 mm
panel, removable 21 mm
bee's wax varnish

panel 19 mm
porous fibreboard 20 mm
woodscobs 150 mm
non-bituminous paper
wood-wool cement boards 35 mm
panel, removable 21 mm
bee's wax varnish

ALTERNATIVE 2
panel 19 mm
porous fibreboard 12 mm
cellulose fibres 100 mm
porous fibreboard 20 mm
panel 19 mm

linseed oil varnish
timber t&g boarding 28 mm
sand 50 mm
non-bituminous paper
wood-wool boards 35 mm
limeplaster

ALTERNATIVE 2
lime-cement plaster
light aggregate blocks
250 mm
wood-wool cement board
100 mm
porous brick 108 mm
water glass varnish

Fig. 4.30   The air transfusive house.

A. Dwelling in Steinkjer, Norway 1988. Architects: Cobolt. Also detail alternative 1.

B. Dwelling in Lørenskog, Norway 1988. Architects: Cobolt. Also detail alternative 2.

C. Dwelling in Stavanger, Norway 1988. Architects: ARK 117, Sandnes. Consultants: Cobolt. Also detail alternative 2.

D. Cow-house in Lillehammer, Norway 1988. Architects: Cobolt. Wood wool boards with air transfusive plaster in walls.

is complicated to gauge. Conventional vapour-sealed construction on the other hand is simple to gauge, because it is undimensional and lifeless.

When a building is suffering from woodworm or rot, is it better to poison the occupants along with the infestation or to use less guaranteed methods that are biologically safer? Organic solvent preservatives give off vapour in the short term; when working on buildings you notice the effects months later, and over many years occupants are exposed to skin contact, and breathe impregnated dust.

Recent studies of woodworkers and do-it-yourself enthusiasts at the Burger Hospital in Stuttgart, West Germany,[5] have proved that toxic agents in both spray and brush applications of timber sealants and varnish, attack the nervous system, with resulting personality changes. The changes take between 10 and 12 years to manifest. Some safer products currently on the market are given in Chapter 7.[3-5]

Diagnosing the tight building syndrome and hypersensitivity to chemicals

Some people are more sensitive to chemicals than are others. In 1963 it was first recognized that chemicals in indoor air could provoke symptoms, but lack of measurements and testing equipment delayed justification of the findings. Because these indoor chemicals exist in outdoor air as well, a more accurate designation might be environmentally induced illness (EI).[6]

Volatile organic hydrocarbons are only a few of the triggers of the tight building syndrome or EI. With the installation of ureafoam formaldehyde insulation (UFFI), a corresponding dramatic increase in symptoms enabled investigation to be undertaken on a wide scale. Formaldehyde became a prototype since this chemical clearly demonstrated a tremendous susceptibility on the part of individuals.

In the years 1984—1990 measurements of ambient levels of formaldehyde have shown that in newly constructed homes and offices, with particle board sub-flooring, panelling, prefabricated walls, clothing, carpets, beds and furniture, high levels of outgassing from this chemical composition have occurred.

Some hospitals had levels of 0.55 μl/l. Formaldehyde, along with a number of other contaminants, is listed in Chapter 7. Occupants of buildings suffer from attacks of headache, nausea, and inability to think, dizziness, lethargy, irregular heart beats, flushing, laryngitis, irritability, unwarranted depression, joint pain, and extreme weakness.[7]

The pioneer work of Dr Sherry Rogers over a period of 10 years at the North East Centre for Environmental Medicine, Syracuse, New York had led doctors throughout the world to re-evaluate medical conclusions on people suffering forms of mental stress from causes unknown. Many of the rules of medicine can well become obsolete when it comes to understanding hypersensitivity to chemicals. The *symptoms* that people are complaining about are unpredictable and variable.

The people born in the 1960s are the first generation to be exposed to buildings that were designed and constructed tight (i.e. with minimal infiltration rates of outside air) and containing an unprecedented number of materials that outgassed (in-built toxins and other pollutants that slowly seep out into the indoor atmosphere).

Dr Sherry Rogers and her co-workers developed a skin-testing procedure[6] to determine means to isolate the toxic triggers. All testing was done single blind, meaning that the patients were unaware of what they were being tested with, and there were many placebo controls (e.g. normal saline or salt water) used liberally throughout the testing sessions so that malingering could be ruled out.

Case studies

1 CP was a 39-year-old consulting engineer who travelled extensively. Two years before his illness he moved into a new house that had new carpeting and particle board sub-flooring throughout. Six months later he experienced an insidious onset of joint pain. He consulted a doctor, but reported that 18 months later he still had the same pains. He also said that he ached more when he was at home.

It was found after consultation with Dr Rogers that his blood serum level of formic acid (a metabolite of formaldehyde) was

10 μg/ml after a weekend at home and 6 μg/ml after a day at work. A passive badge monitor showed that in 24 hours the formaldehyde level in his home was 0.06 μl/l. Single blind testing with normal saline produced no symptoms.

A test dose of formaldehyde produced erythema and weal (a redness and swelling) of 2 mm and the patient complained of ringing in his ears and aching joints. A subsequent dilution cleared or neutralized all his symptoms.

His health improved when he had removed all the particle board and carpeting. He became healthy again, only to find that the old symptoms returned when he visited or stayed in other places with the same pollutant levels.

2   AR was a 56-year-old female who had 11 years of progressively worsening headaches. She could get them anywhere, but they were worse at work where for 15 years she had been employed in the same factory. It was found that after a day at work her blood test level for tetrachloroethylene was 26.1 μl/l. After two weeks at home it had dropped to 18.1 μl/l. She was not free of headaches until her blood had been neutralized and the level had dropped to 1.3 μl/l.

3   DB was a 10-year-old attending school. His teacher sent home reports of falling grades. He was becoming disruptive in class, and unteachable. Single blind testing to common indoor chemicals proved negative, but when phenol was tested he suddenly began to scribble viciously, lay on the floor and kicked the wall. The neutralizing dose terminated the symptoms. The source of phenol proved to be a popular brand of phenol-based cleaning solution the teacher used to wipe down the desks. After this was stopped, the child returned to normal.

4   PS was a 23-year-old female who was in good health until she started a job in a shopping mall. Then she began to experience bizarre behaviour, resulting in long periods in hospital and mental institutions where she was diagnosed as being schizophrenic. She also complained of severe muscle spasms in her neck, and worse when she was at work.

On testing she had no symptoms when given normal saline. A test dose of formaldehyde caused muscle spasms so severe that she could not straighten her neck. She began laughing and rocking the chair and thought she was the wife of Jesus. When her blood serum was neutralized she returned to normal.

From 1984 to 1989 over 1000 patients with undiagnosed chronic symptoms that had not responded to a wide range of normal treatments have been through Dr Rogers' single-blind testing method with chemical additives that were found to be present in the majority of building materials currently being used. There is now definitive and scientific proof that many of the synthetic materials used in modern construction methods are dangerous to health.

The technique of provocation/neutralization of symptoms to diagnose chemical hypersensitivity has now become standard practice for hundreds of physicians in the USA. Some of the chemicals tested have included the more frequently identified indoor chemicals that are difficult to avoid, such as toluene, benzene, xylene, ethanol, trichloroethylene and natural gas.

Increasing the ventilation rate and avoidance of damaging materials in the clinical testing rooms were seen to be pointers for good healthy building design.

Note: It is imperative that the space where any testing is to take place must in itself be free of the chemicals that are being isolated.

The information on Legionnaires' Disease was contributed by Roderic Bunn, and first published in the Building Services Journal *(UK), September 1986.[8]*

Legionnaires' Disease[8]

Since the earliest recognized outbreaks, epidemiological evidence has strongly associated the spread of both the fatal form of the disease, *Legionella pneumonia*, and the non-fatal form, *pontiac fever*, with certain types of domestic and evaporative water systems.

Aerosols containing *Legionella pneumophila* are produced by showers, humidifiers, cooling towers and taps. Many interrelated environmental factors will influence aerosol survival and infectiveness, in particular the effect of ultraviolet light, ozone and bacteria present in the aerosol.

The carry-over of spray from a cooling tower must not be confused with the vapour plume sometimes seen trailing from a cooling tower. The plume is caused by condensation of the vapour which has left the cooling tower as a saturated gas and which in that state would have a particle size too large to inhale.[9]

In towers that are not regularly cleaned, microbial sludge can accumulate in the reservoir, promoting the growth of aquatic organisms including *Legionella*.

Many factors influence the survival of airborne micro-organisms, of which the most important are atmospheric conditions and relative humidity. Broadly speaking, airborne *Legionella pneumophila*, in common with many other bacteria, survive relatively well at medium r.h. values but less well at high r.h. values. Organisms appear not to survive well in dry atmospheres (< 30% r.h.).[10] The relative humidity (r.h.) inside the average building is conducive to *Legionella* survival.

High temperature can reduce the particle size through evaporation and thereby enable the organism to be inhaled. With cooling-tower-generated aerosols, the greater the aerosol age, the more likely that natural biocidal factors in the atmosphere will kill the organism.

Research has shown that oxygen is known to be toxic to many bacterial aerosols, but often the effect may only be apparent at low r.h.; so it is possible that the poor survival of *Legionella pneumophila* in drier air may be due in part to oxygen toxicity.[10]

The graphical information on *Legionella pneumophila* life cycles is derived from a UK Public Health Laboratory Service study (1983).[10]

The growth curve for *Legionella pneumophila* is typical of those for many aquatic bacteria. The information is based on fairly rigorous data and compares with *Legionella pneumophila* life cycles experienced in the field. Therefore they are considered useful as a design tool. More data is on course and should be checked constantly and updated.

A laboratory test[10] in which different samples of *Legionella* were held for 15 minutes at different values of r.h. showed that the organism survived poorly at 55% r.h. compared with its ability at r.h. values slightly above and below this figure: 65% and 40% respectively. It is not known why the 55% r.h. zone of instability exists for the *Legionella pneumophila* bacteria, but it shares this trait with many other bacteria.

At certain humidities, water in the bacteria membrane is lost, causing the membrane to become unstable, thus killing the organism. At lower humidities the water is removed so quickly that the membrane does not have time to become unstable; at higher humidities, the membranes are maintained. This may occur if the *Legionella pneumophila* organism is suspended along with other organisms that will protect it. It is all a matter of interaction between the r.h. and other factors such as inhibitory salts.

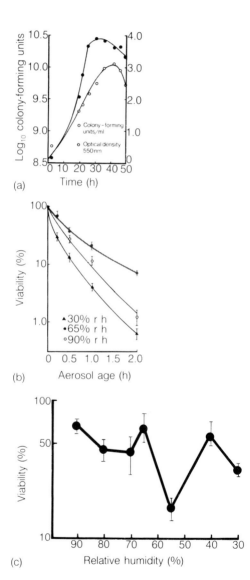

Fig. 4.31 (a) Growth curve of *Legionella* in liquid medium. (b) Survival of aerosolized *Legionella Pneumophila* grown on solid medium and held at different humidities. Standard error, *n* > 4. (c) Survival of aerosolized *Legionella Pneumophila*, sprayed from water and held at different relative humidities for 15 minutes. Standard error, *n* > 4. (Source: *Building Services (CIBS) Journal*.)

Healthy Building Code Proposal 9
Dirty water is one of the main factors in outbreaks of Legionnaires' Disease. Designers, contractors, water treatment engineers, plant managers and maintenance personnel must upgrade their attitudes to care and conservation.

Studies[10] have shown that *Legionella* can retain viability, virulence and aerosol stability when stored in an aqueous environment. The aerosol stability of organisms with low metabolic activity is such that significant numbers of viable organisms can be recovered from bacterial clouds for up to 2 hours.

Since contaminated evaporative condensers might be anticipated to generate continuous aerosols containing viable *Legionella pneumophila* organisms, it is likely that exposed susceptible people might inhale and retain sufficient viable organisms to become infected.

Generally speaking, *Legionella pneumophila* cells in a stationary growth phase, i.e. growing slowly or not at all, tend to survive better in adverse environmental conditions, be it aerosolation or disinfection, than do cells which are going through an exponential phase of growth.

Aerosol age affects viability by allowing the longest action of degrading factors such as temperature, ultraviolet light and gases such as ozone. However, the effect is mitigated by distance from source and continuous generation and exposure.

Particle size determines the rate of deposition, with larger particles settling out quickly. The small-diameter particles (< 5 μm) required for initiation of *Legionella pneumophila* will remain for considerable periods of time.

Dust, micro-organisms and the quality of cleaning

Healthy Building Code Proposal 10
It is the job of every member of the design team to check every aspect of the building environment both in respect of materials of construction, systems of comfort, and ways in which spaces are furnished, to ensure a minimum of indoor pollution.

Internal cleaning of buildings has in many countries become an area of economic cut-backs. The role of the maintenance engineer, once a position of importance in any group of buildings, has been eroded and the skills of repair and reconditioning lost. Instead it has become the realm of the odd-job man or contract maintenance, with replacement of defective parts instead of conservation of building engineering systems by repair, reconditioning, and a respect for the overall building.

Many buildings are becoming sick due to the interaction between dust levels, micro-organisms, the non-upkeep of filters, cleaning of spray nozzles, etc. in ventilation systems, and the quality of cleaning. In addition, the use of synthetic wall-to-wall carpets and other dust-colecting agents has made the job of providing a good indoor air climate almost impossible.

A recent investigation of five Danish Schools[11] was made between the use of carpets, and their removal. Dust combines with bacteria and fungi. The results showed that with fibre carpets the amount of settled dust ranged from 1200 mg/m^2 to 230 mg/m^2 for a reasonably newly laid carpet. Good-quality linoleum gave a low reading of 65 mg/m^2.

Since the discovery of house-dust mites in 1964, the association between house-dust-mite allergy and asthma has been the subject of intensive investigation. Jens Korsgaard of the Chest Clinic, Aarhus Municipal Hospital, Denmark[11] has concluded that the most important factor leading to growth of house-dust mites in all buildings is high indoor humidity, and that the single most important factor for the development of mite-induced asthma is a high exposure in bad, humid habitation.

Passive smoking:

	Particulate concentration (μg/m^3)
Dwellings with no tobacco smoke	91
Dwellings with 0.5−10 cigarettes/day	169
Dwellings with > 10.0 cigarettes/day	475

Epidemiological studies clearly show that passive smoking can create a small decrease in lung function. But evidence on the development of childhood asthma and aggravation of respiratory symptoms in adults was still conflicting in 1990.[11]

1 A level of 100 house-dust mites per gram of dust must be regarded as a risk point for the development of asthma in those people susceptible.

2 A level of 7.0 g/kg is the level of absolute humidity above which excess mite growth will occur. (In some geographical areas a level of indoor humidity below 7.0 g/kg can be considered.)

One of the problems in the design of healthy buildings is that different elements in the cocktail of pollutants require different design criteria for their control when dealt with in isolation. Thus Korsgaard in his clinical interpretations suggests that minimum ventilation air rates are provided to control dust-mite generation. But this is in direct contrast to the need to provide increased ventilation rates to reduce other pollutants.

This means that materials and humidities that provide home comforts for dust mites should be reduced to acceptable limits. Control methods should include:

1 Thorough cleaning of carpets, or complete removal, reduces fungi and dust mites.
2 An inexpensive and labour-saving method is to seal rooms and to allow evaporation from an open bowl containing a mixture of 1:1 vinegar and clorox for 48 hours, with 48 hours' clearance time (50% success rate).
3 Installation of an electrostatic precipitator showed after 1500 hours' operation a 50−100% reduction of fungi.

Dr Sherry Rogers also found in her work that increasing ventilation raised the ambient level of the fungi, which agrees with the work of Korsgaard. Outdoor fungi, like the dust mite, find indoor temperatures and humidity more conducive to growth. This points to a need for heat-exchange units and filtration at all points of entrance to the room, since increased ventilation is one of the major ways of reducing ambient indoor chemicals.

Healthy Building Code Proposal 11
Collaboration between the architect, HVAC engineers and specialists in environmental medicine is essential.

The olf and the decipol

New units of measurement for indoor pollutants and comfort equations for indoor air quality and ventilation are being established. Professor P.O. Fanger of the Technical University of Denmark[12] introduced the new units of environmental measurement, the olf and the decipol. These new units for air quality are analogous units for light and noise.

The symptoms that have now been termed the sick building syndrome have been well documented in hundreds of detailed field studies in offices, schools, dwelling houses, and other non-industrial buildings in many parts of the world.

Although, as Fanger found, the quality of the air supply may for instance be lower than designed for, the frustrating fact is that most of the buildings studied in different parts of the world complied with existing ventilation standards, and yet up to 60% of the occupants found the quality of the air unacceptable.

The purpose of a ventilation standard is to provide acceptable air for the occupant. A comfort equation for air quality considers all the pollution sources present in a space, in the same way as they are perceived by people (by means of the 'sniff factor').

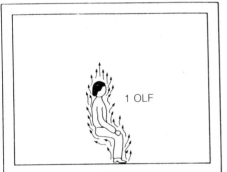

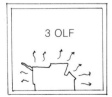 

Fig. 4.32 (a) One OLF is the air pollution from one 'standard' person (an average adult working in an office or similar non-industrial workspace, sedentary and in thermal comfort with a hypienic standard equivalent to 0.7 baths/day). (b) A pollution source, e.g. laser printer has a strength of 3 OLFs if the pollution from three 'standard' persons causes the same dissatisfaction as the source.

The olf

One olf is the air pollution from one standard person (an average healthy adult who takes a shower or bath a day, working in a sedentary situation in thermal comfort). See Table 4.4. A pollution source can have the strength of n olfs, if the pollution is the same as that from n standard people.

Table 4.4   Olf values for human pollution sources

	Metabolic classification	Olf value
Sedentary person	1	1
Active person	4	5
Active person	6	11
Smoker (when smoking)		25
Smoker average		6

Isolated materials pollution

Chipboard	<	2.4	decipols
Synthetic carpet	<	3.4	
Lacquer	<	3.7	
Tobacco smoke	<	14.4	

It should be noted that the 'smell' (olf/decipol) factor does not always relate to the latent effect of any outgassing upon a human being.

Comparison of ventilation comfort standards

Comfort equation (Franger) (litres/sec)/m^2
IAQ = 1.4 decipols
      Existing building
        0.7 olf/m^2 smoking 5
      Low olf building
        0.2 olf/m^2 no smoking 1.4

ASHRAE Standard 62–81
      Smoking   1.7
      Non-smoking   0.25

Nordic guidelines
      Smoking   1.0
      Non-smoking   0.4

DIN 1946 standard
      Smoking (large
        offices)   1.9
      Non-smoking
        (large offices)   1.4

The curve for the dissatisfaction caused by one olf at different rates of ventilation (where dissatisfaction is expressed by people who find the air unacceptable when entering the space) is based on bioeffluents from more than 1000 people judged by 168 investigators.[12]

The decipol

The concentration of air pollution depends on the pollution source and the dilution caused by ventilation. The perceived air pollution is defined as that concentration of human bioeffluents that would cause the same dissatisfaction as the actual air pollution.

One decipol = 1 olf ventilated by 10 litres/sec of unpolluted air

The percentage of dissatisfied occupants is given as a function of the perceived air pollution (decipols).

The decipol level expresses perception of the air by people, not whether the pollution is a health risk. Such risks have to be considered separately. But as Fanger rightly surmises, our senses — with a few exceptions — are also influenced by harmful pollutants.

Thus a decipol scale/person can be drawn.

The comfort equation for IAQ and ventilation is:

$$C_i \ = \ C_o \ + \ 10 \ G/Q$$

where C_i is the perceived air pollution in space (decipols), C_o is the outdoor perceived air pollution (decipols), G is the pollution source strength in space and ventilation system (olfs) and Q is the outdoor air supply (litres/sec).

Professor Ole Fanger found from an investigation of 15 office buildings in Copenhagen that the average pollution source was 138 olfs.[13]

Pd = 395 · exp (−1.83q$^{0.25}$) for q ≥ 0.32 l/sec. olf
Pd = 100%   for q < 0.32 l/sec. olf

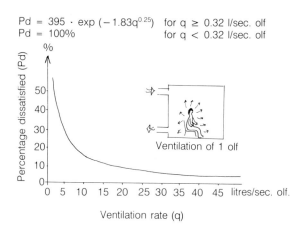

Ventilation of 1 olf

Fig. 4.33 Dissatisfaction caused by 1 OLF at different ventilation rates. Those dissatisfied are the people who find the air unacceptable when entering a space. The curve is based on Professor Fanger's experiment of the bioeffluents from more than one thousand persons, judged by 168 subjects.

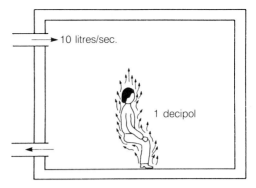

Fig. 4.34 One decipol is the perceived air pollution in a space with a pollution source of 1 OLF ventilated by 10 litres/second of unpolluted air. Steady state conditions and complete mixing are assumed.

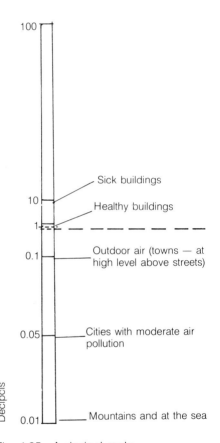

Fig. 4.35 A decipol scale.

Indoor air quality (IAQ)

Volatile organic compounds (VOCs)

Researchers at the US Environmental Protection Agency (EPA) have detected upward of 300 compounds in a single building, and have identified a total of over 900 separate compounds in indoor air. In Denmark 62 frequently used chemicals were found emitted from 42 different building materials, and 84% were known or suspected mucous membrane irritants (see Table 4.5).[14]

This information could explain why there is such a high rate for irritation of the nose and throat in sick buildings. Even when irritation and illness complaints subside, the fact that some 28% of the compounds identified were suspected carcinogens is reason for substantial concern about the materials that we are currently using in building construction.

From a study in a Swedish nursery school in 1988[15] it was found that the building materials appear to act as a sponge for VOCs, taking contaminants in, and then releasing them later, often in a new chemical form, since they can mix with indigenous pollutants already in the material (e.g. blue concrete, an alum shale-based aerated concrete, has a high uranium/radium content of 700−2500 Bq/kg). The conclusions drawn from this study are that:

1 Much of the initially released VOCs will be absorbed on to other material surfaces from new furnishings or finishes in spaces containing high-surface-area materials, such as carpeting, ceiling tiles, or free-standing partitions.
2 The quantities absorbed will depend on the total surface area exposed as well as on the air exchange rate in the space. The rougher surfaces of insulation materials, textiles and carpets will absorb large quantities of VOCs. The available surface area for absorption of fleecy materials is many times the plane surface measurement.

Table 4.5 Health effects of 62 organic chemicals emitted from 42 building materials

Effect	Mucous membrane eye irritants (%)	Carcinogens (%)
Known	48	nil
Suspected	36	28
Unknown	1	72
None	15	nil

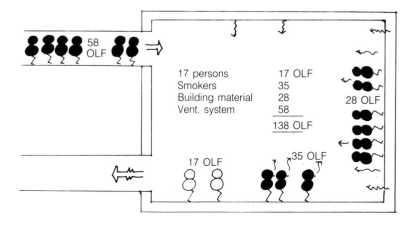

Fig. 4.36 Average pollutant source in 15 offices, Copenhagen, Denmark. The average workforce was 17 people per office. The many hidden emissions (out gassing) plus smoking are the probable cause to the mystery of sick building.

17 persons	17 OLF
Smokers	35
Building material	28
Vent. system	58
	138 OLF

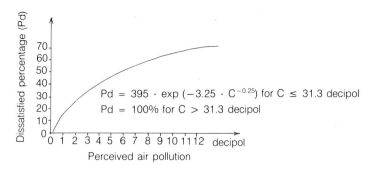

$$Pd = 395 \cdot \exp(-3.25 \cdot C^{-0.25}) \text{ for } C \le 31.3 \text{ decipol}$$
$$Pd = 100\% \text{ for } C > 31.3 \text{ decipol}$$

Fig. 4.37 Percentage of dissatisfied as a function of the perceived air pollution in decipols (derived from data in Fig. 4.33).

3 The lower the ventilation rate, the more VOCs will be absorbed on interior surfaces. When ventilation is reduced or turned off (at night and at weekends or during periods of warm or cold outdoor temperatures) as well as restriction to reduce high input of external pollutants (see Section 4.1) indoor air VOC concentrations will tend towards equilibrium; i.e. they will rise until there is a balance between emissions from sources and removal from the air by ventilation, chemical reaction or surface absorption. As VOCs are emitted materials they will be absorbed on surfaces.

4 The higher the temperature, the higher the emissions and VOC air concentrations (see p. 93 for the case study appraisal).

Healthy Building Code Proposal 12
- Use maximum outside air ventilation during and following the installation of finishes and furnishings to reduce levels of VOCs emitted from new products and materials that are not free from such compounds.
- Use temporary exhausts (through doors, windows that can be opened, stair towers and emergency exists) for exhausting air, rather than the HVAC return system wherever possible. It is important to operate ventilation systems 24 hours a day, 7 days a week during periods of elevated VOC levels.
- Protect installed materials with vapour barriers where feasible during the use of finishing products with high VOC concentrations (e.g. adhesives and paints) and during installation of VOC-emitting furnishings and partitions.
- Operate newly occupied building areas at the lowest temperature acceptable to occupants. Temperature excursions can cause bursts of VOC concentrates that have low boiling points.
- Avoid the use of fibre-lined HVAC ducts. Avoid the use of spray-on insulation (unless proved to be low on outgassing) for return air spaces and acoustic chambers.
- If possible do not use materials with high VOC factors (see Chapter 7).]

Healthy HVAC systems design

Indoor air quality problems will persist and increase, both in new buildings and in retrofitted properties. The building services engineer/building physicist is at the sharp end of this particular building function. He must constantly update his knowledge of the sort of problems that IAQ can produce.

It is important that this book and other guides are seen as primers for seeking further information that must be constantly updated. When conditions are outside the designer's control, due to economics or other constraints, he/she must be careful to document their design conclusions and thought patterns, and be sure to transmit such information to the responsible parties — and in particular to the client. This is of course necessary at all stages of a building design by every person in the team where the health of a future occupant can be at risk.

Sealants are ineffective for PCP control in timber-framed buildings

Where wood is used in the construction of timber-framed homes, and as a decorative internal finish, treatment with PCP as a wood preservative has been found to have a high toxicity which can affect people, particularly children, according to the US Environmental Protection Agency (EPA). The EPA[16] has also shown that sealing the treated wood has little effect on PCP levels in indoor air.

PCP's toxicity is attributed to the common chemical formulations: dioxins, furans and hexachlorobenzene. Commercial PCP is a teratogen in laboratory animals and a possible animal carcinogen. The US National Toxicology Programme has found clear evidence of carcinogenicity in laboratory mice.

Because PCP has a relatively low vapour pressure, it tends to be emitted very slowly from materials that have been treated with it. PCP can still be found in building air 15−20 years after initial application.

Studies have shown that as a toxic element it can enter the human blood and urinary systems. The highest levels are found in children. The EPA report stresses the importance of higher ventilation levels in buildings.

The formaldehyde story

ASHRAE reviews its codes in a 5-year cycle. Code 62-1973 was reissued in 1981 with a new approach to ventilation air rates. The minimum rate of 5 (ft^3/min)/person (2.5 (l/sec)/person) was kept. However, because it was known that air quality was becoming lower, innovative solutions were proposed to increase the rate of ventilation in areas where higher concentrations of outgassing were found. Hardly had the ink set, when the Code was revised, whereby the minimum value was tripled. But even before that the Formaldehyde Institute had objected to the standard.[17]

Due to increasing energy costs and conservation, buildings are being built with tighter envelopes; yet buildings air-leak more than expected. Research has shown that infiltration is often above one air change per hour. In many cases the infiltration alone can meet the ventilation requirements of the building. However, if this infiltration can be allowed to flow through a labyrinth within the building structure then the normal energy additions can be reduced (see Chapter 5 for examples).

Unfortunately, reduced fresh air to buildings due to specifying below the minimum standards has led to mechanical ventilation systems not having sufficient air quantities to deal with the outgassing found in many modern building products and items of internal machinery. It is often forgotten by designers that even the minimum air quantity given has been based on extra clean environments rarely found in practice.

Control methods

Source removal   Source removal or substitution is the most effective means, but difficult when the source is sometimes difficult to identify. This is of particular importance when dealing with existing buildings and refurbishment.

Source reduction   Source reduction by the use of enclosures, coatings, etc. is usually expensive or ineffective (as, for example, with sealants for wood treated with pentachlorophenol (PCP)).

Behaviour changes   Behaviour changes which involve personal choices by occupants in their use of space, and methods to minimize risks from exposure to machines that pollute, are often seen as real or perceived restrictions on lifestyle.

Ventilation methods   Ventilation methods open to the HVAC designer are many. The use of systems normally reserved for industrial removal of contaminated air sources can, with degrees of sophistication, be adapted for commercial and residential use.

Designer tools

Codes and standards   Codes are the regulatory means to protect the public from IAQ problems and other hazards. Most countries have sets of codes. In the USA these are based on the ASHRAE Ventilation Standard 62. (ASHRAE codes are referred to more than others, since many countries in the world without established standards of their own use the work of ASHRAE as their guide.) Standard 62 is the only consensus standard available in the USA, and is therefore considered the source of state-of-the-art information.[17]

Like all other ASHRAE standards, ASHRAE 62 is advisory and has no force in law. Codes are intended as references for adoption by others where appropriate. As with the CIBSE in the UK, ASHRAE standards have been established for over 50 years, so they tend to be taken as the 'final word'. Like all guides, many of these recommendations which worked for much of the time are now found to be inadequate, due to a gradual change in the quality of indoor air, which has begun to give all designers second thoughts.

Rapid advances in our understanding of indoor pollutants and their interaction will mean that all Codes of Practice will need constant updating. Manufacturers will complain, even where

Investigation undertaken by the University of Agriculture, Wageningen, The Netherlands (LUW) in 61 offices[18]

Health complaints	Average (%)	Highest (%)
Eye	19.5	39.5
Nose/throat	23.5	45.1
Neurological	20.3	51.3
Fever	8.8	33.3
Climate complaints:		
Temperature	54.6	89.3
Air quality	45.7	82.3
Lighting	30.0	53.3
Dry air	43.5	80.2
Other complaints:		
Electric shocks (static)	10.5	35.5
Dirty taste in mouth	7.2	20.5
Noise	25.1	50.0

The investigation found that many of the complaints resulted from poor design, poor control, and failure to keep the building services installations in good repair.

the Codes of Practice are not mandatory.

ASHRAE 62-1981 is currently listed as the state-of-the-art code to be worked to:

1 Minimum rate of ventilation is 15 (ft^3/min)/person (7.5 (l/sec)/person).
2 Increased values for areas where smoking is permitted are given.

In the light of the work of Professor Fanger in Copenhagen with the introduction of the units olf and decipol and the work of researchers at the Dutch National Building Research Centre in 's-Gravenhage (see Chapter 8), the whole approach to the use of ventilation rates will become part of a new mathematics of chemical and bacterial pollution. Because of the many variables which are not under the control of the HVAC designer there is no final solution to IAQ problems. As with thermal and acoustic standards, designers must walk a careful path, evaluating every aspect of the proposed design, making themselves as informed as possible on every element that is being knitted together, and then trying to change the material being used, or indeed the whole design of the building.

Important pollutants: comfort and health effects

Comfort   Odours are by far the most significant and common IAQ complaint. They can affect an occupant's performance or attendance at work. Comfort is restored by leaving the polluted area.

Acute health effects   These effects can be short-term diseases such as colds, influenza, asthma, etc. They cause absence from work and usually clear up once the offending material has been removed.

Chronic health effects   These are long term and irreversible, sometimes taking decades to develop. Lung cancer is an example. Usually without sensory perceptions of the hazard, the occupants are not immediately aware of it.

Updating information

All design offices should invest in an electronic library, where information can be quickly updated and referred to at the touch of a button during design team meetings. Such information should become internationally integrated. The research being undertaken in different countries in the world must become common property for a better understanding of the problems. Also, individual countries should themselves, at governmental level, investigate all new materials manufactured and codify both the energy content and the contamination content.

Other reviews by the designer should be system design concepts, system operation and system maintenance.

Many modern HVAC systems lack the essential installation of cleaning doors, and space to maintain without hindrance. The use of high-velocity small-tubed ducting has its advantages, but making equipment smaller to answer the ever-increasing pressure of architects to 'hide away' the services has resulted in the use of materials that accumulate dust and moisture and become hazard niches for the rapid growth of micro-organisms, moulds, etc. These materials become health hazards for the occupants. Built-in warm-air systems spell disaster for many occupants. The materials used are often damaged during construction. Return-air systems with no fresh-air connection and filters that are poor in quality, not cleaned frequently, or just left out, create a recycling of all the indoor contaminants.

Checks

1 All air filters must be checked for their material content. An important point is that particulate air cleaners are vulnerable to becoming pollutant traps if not kept clean (refer to ASHRAE code 52-1981 and updates).
2 No standards exist (March 1990) for the performance in ventilating systems of gaseous removal equipment. All manufacturers' data must be critically evaluated.
3 Heat-reclaim methods are important for energy conservation. All methods must be critically evaluated for impact on IAQ.
4 The introduction of local exhaust systems and differential pressure control for pollutant migration control in spaces should become a designer tool.
5 Variable air volume (VAV) systems are a popular energy-conserving technology. They require special attention in respect of IAQ since it is difficult to ensure that the correct rate of outside air is delivered to all zones of the building under all operating conditions. Different occupancy loads and other systems have several VAV boxes on each air-handling system. However, manufacturers in the UK, Sweden and the USA are aware of this, and are using in-built computer sensing of all sides of the equation to bring new products on stream.

Prescription for a healthy HVAC system

1 Work with all members of the team.
2 Establish best estimates of planned and future uses of spaces and their occupancy and process loads.
3 Investigate materials for both building and furnishing for outgassing factors. Refer to current codified information as supplied by manufacturers (obtain from suppliers documented guarantees).
4 Check all building codes and reconcile differences.
5 Use the minimum fresh-air rate of 7.5 (l/sec) per person. But if the ventilation effectiveness is expected to be much less than 1.0 increase the fresh rates.
6 Minimize building exhaust recycling into intake locations which can induce pollutant intake from loading docks, kitchens and other contaminating areas.
7 Consider local exhausts for special situations near a concentration of polluting machines and introduce differential pressure control to prevent contamination flow.
8 Provide means to increase or decrease fresh-air rates as space changes occur, making flexibility of design a positive control mechanism in the healthy building design process.
9 Check all heat-reclaim equipment.
10 Where appropriate install particulate filters on all return-air points. Use gaseous removal equipment.
11 Select materials that do not produce or promote contamination. Non-metallic duct linings can produce airborne particles, collect dust and bio-organisms.
12 Avoid standing water in condensate drains, equipment rooms and ducts. Be sure that condensate drains are dry. Standing water has been implicated in many IAQ problems since some airborne particles lodge there. Under certain temperature

ASHRAE Research and Technology Committee are currently (March 1990) finding replacements for harmful chlorofluorocarbons in refrigerating machinery with the support of the US government; improving indoor air quality by developing basic data on air leakage of selected building components; completing full-scale testing in commercial office buildings to gather data on IAQ, particularly the presence of volatile organic compounds (VOC); evaluting the efficacy of certain biocides against Legionella *in cooling towers, as well as studying human response to stimulated air movement in rooms.[19]*

Tests by Danish scientists on international brands of photocopiers and laser printers have found that after only 1 year the machines release high levels of ozone gas ten times above the UK safety standard. Even the threshold of 0.1 part per million is enough to cause ear, nose and throat irritation, while levels five times as high can lead to nausea, headaches and increased risk of lung infection. Scientists at the Danish Technological Institute in Copenhagen, while testing 200 offices, found internal filters that were clogging up and had stopped working after 12—18 months, producing emission of up to 1 part per million.

The risk of ozone pollution is at its highest in small, badly ventilated offices where it is difficult for the gas to break down. Ozone from printer fans can spread up to 5 m around the machine. In Denmark firms are now banned from placing printers in normal work places. Supplementary filters on laser printers have been made mandatory.

Special rooms should be provided with industrial air change rates by mechanical means with stand-by operation. Six to ten air changes per hour would be appropriate.[20]

conditions rapid growth can occur and large quantities of bio-matter can be entrained in the duct system (condensate pans in window air-conditioners have caused problems). Where sumps are necessary, procedures for cleaning the water should be provided.

13 Provision of easy access for equipment maintenance and cleaning of ductwork is a must.
14 Easy access must be provided for testing and checking instruments for system commissioning (see Chapter 6).

Documentation

It cannot be stressed too much that high standards of commissioning and documentation are both important to ensuring that a building once complete is as healthy as possible; and that the HVAC designer and the rest of design team as well as the installing contractor should be protected from future legal action if problems arise, especially if during the design process 'risk' decisions have been made.

Do not be put off by anyone in the execution of quality commissioning and control, whether it is the client or the contractor. It may be hard when one is pressurized to get the building occupied but it allows you to rest better at a later date.

1 Include records of all design choices and the rationale behind them. This is particularly important if constraints on the project have dictated changes against your firm advice.
2 Include complete operational procedures to ensure that operation is as the design requirement, even if these are not your responsibility.
3 Ensure that your commissioning section relates to all items of work that can create pollution problems from whatever source.

Quality improvements with displaced ventilation

Older buildings of solid construction, high ceilings and minimal internal heat loads result in small temperature variations both during the day and throughout the year. Such buildings are often not mechanically ventilated or air conditioned, yet the occupants find such buildings good to live and work in.

Technical solutions since Roman times have been based on fundamental physical laws. Simple solutions to finding a comfort environment were obtained where the builder worked with natural laws. Air was allowed to be warmed under the floor and then to rise, thus displacing the room air which eventually became warm, and contaminated. The air was changed by displacement.

We have seen from other sources (Chapter 5) that to make buildings healthy we have to flush the buildings through with air that is as clean as possible, with the absolute minimum of recirculation. No longer are we just concerned with the thermal aspects of indoor climate control; we have to contend with the worrying concentration of room contaminants. Displacement ventilation provides considerably better quality air than circulatory systems.

Terminology

With displacement ventilation air is delivered to a room at low velocity. Professor P.O. Fanger advises sensitive velocities of

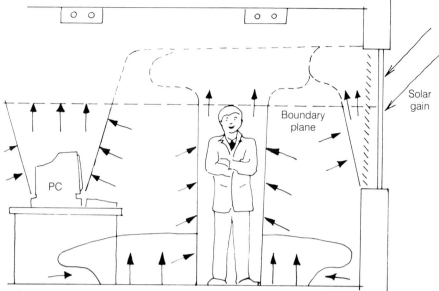

Fig. 4.38 Bouyancy assisted room air diffusion, heat sources and plumes in a typical office. (Source: *Building Services Journal*.)

0.80–0.1 m/s. The more accurate term is buoyancy-assisted mechanical ventilation, giving weight to the main driving force.

Air terminal devices

There are two main approaches for the supply-air devices used with buoyancy-assisted mechanical ventilation and air-conditioning:

1 Supply air at or near floor level, via a low velocity device.
2 Supply air in the form of rapidly diffusing jets via floor, desk of seat-back devices. (Many air terminals will come on to the market, and point of extraction devices for high-risk pollution zones should also be independently designed to suit specific applications.)

Low-velocity floor terminals have been widely adopted in Scandinavian countries. Manufacturers include Stratos, Stifab, Farex and Halton. Floor, desk, and seat-back devices were primarily developed in West Germany by the manufacturer Krantz, now joined in the UK by Trox and Waterloo.

Design procedure

The choice between high and low-level supply has a major impact on many aspects of the design process, not just the ventilation or air-conditioning system but also of the building and its internal organization. The decision is complex but will usually be based on a combination of:

1 Economics, i.e. total cost to the user
2 Aesthetics
3 Space: use of false ceiling, floor void, wall recesses or floor area
4 Comfort: potential for draught or excessive temperature gradient
5 Hygiene: mixing versus displacement. Question: Is there a contamination problem?
6 Potential for heat recovery
7 How heat losses (if any) are to be dealt with

The approach to design depends on whether the dominant need is contamination or temperature control. Whereas contaminant control dominated industrial process ventilation, we are beginning to see that we have to find a balance in the way we design for commercial and non-industrial use. The reference to the high rates of contamination from office copy printers is a prime example of treating the matter of space ventilation in a way where the risk elements are seriously judged.

Contaminant control

In true displacement ventilation, contaminants are pushed ahead of a piston of pure air. If the process is to be stable, and not influenced by forces such as buoyancy, gravity or moving bodies, then there must be sufficient momentum in the air to withstand these forces. This normally requires a high and even velocity (0.4 m/sec) across the whole room, leading to air-change rates of 500/h.

A good example of this method of ventilation is the laminar-downflow clean room.

With buoyancy-assisted ventilation, contaminants are usually released at, near, or by the heat sources, and carried out of the occupied zone in the plume. As stated before, semi-industrial processes (copy printers, medical test booths, smoke booths, etc.) require local exhaust ventilation.

Note: The volume of air Q (m^3/sec) required to produce a reduction in concentration in the occupied zone C_r (ppm) of a known contaminant being released into the occupied zone at a rate of q (m^3/sec) is calculated from:

$$Q = q \times 10^6/(C_r \times E)$$

where E is the ventilation index factor which increases with the thermal energy available in the plumes and the height of the room.

There is very little published guidance on how to determine a value for E. In theory, values can be as high as 10, but measurements so far indicate that a reasonable design maximum is about 2, with 1.4 to 1.7 being more usual.

Conclusions

Buoyancy-assisted ventilation with the use of low-velocity air diffusion devices is very suitable for tall spaces with non-sedentary activities and high-temperature contaminating processes. It is difficult to design for the office application, but the potential for improved air purity and lack of draught around the neck region may well be worth pursuing. Furthermore, elevated air temperatures create good conditions for air-to-air heat recovery even in moderate climates.

Both the low-velocity induction device and the rapid-diffusion floor-mounted air terminals offer particular advantages for commercial applications, with moderate heat gains, the former being particularly suited to buildings which cannot entertain a raised floor. Terminal-generated noise tends to be lower.

Note: Workplace HVAC is being proposed as the answer to dealing with both the individual differences in thermal requirements, and contaminants from source points of machines such as copy printers etc. The micro-climate of each workspace can be more easily optimized and placed under individual control. In such cases, sub-floors in buildings (i.e. offices) can be used for point control of clean tempered air for breathing, and for power and control cables for point-of-use services.

Notes on ventilation systems

1 Rotary air-to-air heat exchangers must be checked for their ability to carry over pollutants.

2 All ductwork systems should be kept clean. (There is a need for manufacturers to market systems of internal duct cleaning.)

3 Return air can be used provided that self-cleaning filters are installed from outlets to all rooms. (However, this should be used only where a thorough search has been made to keep contaminating materials to a minimum.)

Aspects of thermal comfort and room convection fields are well covered in an article entitled 'Design guide for displacement ventilation' by Paul Appleby.[21]

Electromagnetic radiation (EMR) plus some other factors producing the 'cocktail effect'_

Magnetic fields originating from man-made sources are many times greater than naturally occurring fields, particularly at frequencies of 50−60 hertz (Hz), at which domestic electricity supplies operate. In 1969 the World Health Organisation in its report, 'Magnetic fields environmental health criteria',[22] indicated an increased risk of cancer with exposure to very weak fields of 1−10 milliGauss (mG). Such an opinion cannot be dismissed, although actual evidence of cancer is still hard to come by. Like many other 'factors of influence' it may well be that the greatest danger is from a conbination of individual sources, each having some factor of risk — but considered at this present time to be low and 'acceptable'. Put together they could provide a lethal cocktail that may prove hard to analyse.

In a study comparing four Australian houses, Heather Robertson of the Department of Architecture, Royal Institute of Technology, Stockholm,[23] found no support for the popular assumption that pylons near houses mean high internal measurements. Other factors such as house wiring and electrical appliances within the home could have had a greater contribution. Also, many other criteria for healthiness, such as pesticide history, dampness and the emission of fumes from poorly maintained oil-fired furnace heaters, needed consideration. If there were an interaction between such factors, the concept of a 'minimum allowable level' of exposure for each category of pollutant would be inadequate to define a healthy building or a healthy site.

Evidence is beginning to emerge that high exposure to radar, radiation from television masts, television sets, VDUs and even the kitchen micro-wave is causing metabolic malfunctions, skin cancer and even miscarriages.

A recent statement in Britain by the Radiation Protection Unit that high intensity house lamps could by direct exposure to the skin cause skin cancer shows that we have to be prepared to put more finance behind the research into electro magnetic radiation and the interactive effects of all other pollutants.

Electro-biological research is beginning to show how significant the effects on health are. To minimize the effects, building should be located well away from transformers, overhead electric railway lines and power transmission lines. Where there is a shortage of land due to high population densities, as witnessed in The Netherlands, this is a difficult task to perform. New electrical distribution networks are taking over existing urban areas. As the linear conurbations of cities become intensified both in the urban west of Holland and now flowing east−west, between the Rivers Maas and Waal, the demand for remotely-created electricity increases. The factor of electrical pollution is the same, whether the electricity is generated benignly by wind turbines

Electrical diseases:
These now include miscarriages, and metabolic malfunctions from high exposure to radar, radiation from television sets, VDUs, microwave ovens, etc.

In the USA 15% of cancers in children are being caused by exposure to electromagnetic fields. Visual display units (VDUs) increase the risk of miscarriage of birth by 80%.
Environmental Digest, *no. 14, August 1987*

in coastal regions or generated by natural gas, coal or nuclear powered generators.

The emphasis in such built environments is for protection to be afforded within the building itself. Spur cables and electrical services in metal ducts should be screened with thick loam or clay, and routed around areas where the daily occupancy is at a minimum. There should be a distance of at least 1000 mm of electrical cables from water pipes, sleeping and sedentary working positions. Sensitive zones, such as bedrooms should have the facility to be switched off on demand. The houses designed by the Dutch architect Renz Pijnenborgh at Den Bosch, as described in Chapter 5, provide further practical information on ways of reducing the harmful effects of electro-magnetic force fields.

Health risks associated with electromagnetic fields

There are many devices that we use daily in our homes and places of work that have an immediate risk/benefit ratio that we can deal with individually and in the way we plan our buildings.

The ambient field

The level of field strength that we are constantly exposed to is produced by local electric-power transmission and distribution networks, as discussed in the Australian example. This field is present both outside and inside our buildings. Shielding our buildings from the ambient field is a practical impossibility.

In urban environments the ambient field often exceeds 3 mG, whereas in rural areas it is generally less than 1 mG; however, these numbers may be higher, depending on the proximity of electric-power transmission lines, transformers, etc. The best evidence currently available[24] is to take 1 mG as the maximum field strength for constant exposure to 60 Hz fields.

Fig. 4.39 The range of frequencies related to electrical apparatus and equipment (after Netherlands Institute for Building Biology and Ecology).

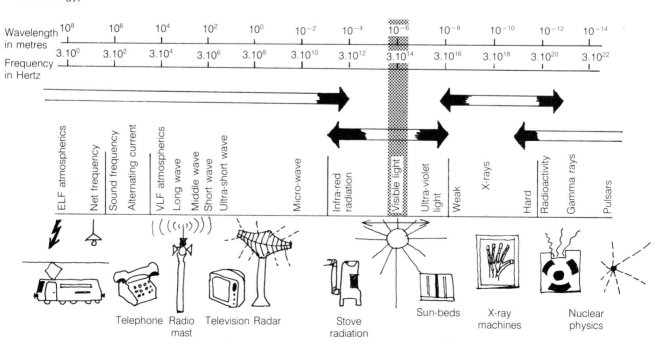

Dose rates and common articles in daily use

Electric razors and electric blankets A line-operated (i.e. plugged into a wall socket) electric razor can have a 60 Hz field of 200−400 mG, 15 mm from the cutting edge. With recent knowledge that a 60 Hz field of 3 mG has been shown to be significantly related to increases in cancer rates[25] the use of such devices even for a short period becomes questionable.

It is the period of use that must be related to the field force. The electric razor is used for a few minutes perhaps once a day, so the dose to the user is minimal, unless he or she is a hairdresser. Electric blankets have a much lower field strength of 50−100 mG, but the blanket is placed close to the skin and used for much longer periods.

The incidence of miscarriage has been found to be much higher among pregnant women who used electric blankets than among those who did not.[26]

The answer is the simple action of heating a bed before one retires, and then switching off and removing the plug before sleeping. The same common-sense approach is required for other devices that give off electromagnetic fields.

Television sets A small amount of ionizing radiation is given off by the television screen as X-rays. The television set is also a broad-band radiating source, giving off frequencies ranging from 60 Hz power frequency to radio frequencies in the megahertz range.

In rats exposed to commercial television sets 300 mm from the animals for 4 hours a day over a period of 60 days the foetal weight was reduced and the functioning of the rat's brain was affected.
Institute of Occupational Medicine, Lodz, Poland, presented to a conference in Stockholm 1987.

The majority of sets on the market working in the VLF range give out radiation levels of over 1 mG, 1500 mm from the front and rear of the set, and only 260 mm to 800 mm from the sides. It is important to understand that electromagnetic radiation passes through wood and other usual building materials. Consequently, when a television set is placed close to an adjoining bedroom, the radiation will pass through and affect a sleeping person or child. Television sets should never be placed in such positions. The same problems exist with other display units.

Personal computers and video-display terminals (VDTs) Radiation patterns here are similar to those from television sets; the major difference in the hazards associated with use is that compared to the television viewer the computer operator must ordinarily sit closer to the device. The field pattern of a Macintosh Plus computer that meets present leakage standards of 1 mG is between 900 mm and 1.5 mm.

The important difference between television sets and PCWs is that the radiation given off by the computer contains the information that the computer is processing. It is quite possible to read this information from half a mile away.

Such high levels of radiation have proved to be the cause of miscarriages. The numbers of clusters of miscarriages continue to multiply.[27] Many computer models have attached keyboards, so that the operator's head is only 400 mm away from the field source. Because the field force drops off with the square of the distance, the need is to work at least 1 m from a machine, which will mean that office and workplace furniture must be designed to suit these new risk elements, and that office space must also be adequate, thus placing a new premium on space requirements for modern office buildings. (I tried to type this book on my Amstrad when at least 900 mm away from the screen, but often with difficulty.)

Computer use should be limited to short periods. This is of importance in schools, but also in offices. I have found that I have been in front of a computer screen for over 8 hours a day. The exposure risk of computers has yet to be codified.

Internal electrical services in buildings

The introduction of shielding devices for computers to reduce the electrostatic field that is emitted from the screen has claimed some reduction, thus reducing the possibilities of complaints such as headaches, skin rashes, and eye discomfort. The projection of a picture onto the surface of a screen calls for a voltage as high as 8,000 to 20,000 volts. In the case of colour sets the voltage is much higher. An electrostatic field tends to leak out towards the user. Even we as users have our own electrostatic field. Low relative humidities can contribute to increased static charges of the human body. The air around us is filled with many different types of particles with either negative or positive charges and the electrostatic field existing between the screen and user affects their movement in the air. This causes the charged impurities in the air to be drawn to the screen or the user. Most of these charges have a positive field and are drawn to the negative field that we as human beings tend to have, thus as much as 10,000 particles per sq mm of our skin is covered per hour.

Recent tests on a simple Swedish invention using a screen and keyboard discharge conductor attached to an earth point and the wiping of the screen twice a month with an anti-static liquid has resulted in tests by the British Standards Institute giving results of a 95 per cent reduction in the electrostatic field. These findings have been confirmed by independent testing by the Swedish National Institute of Radiation Control SSI 84—08 and the Swedish Office Employee Organization.

The tests on the SCREENSAFE device at the BSI were undertaken in the early spring of 1990, and issued on report No. 173750 dated 26th June 1990 (see p. 137).

A common precaution when designing electrical cable ducts and service tracks is to ensure by special forms of insulation that the electrical field forces from a high-voltage source do not interfere with telephone and other communications systems. A standard enough procedure, but rarely does the designer in ensuring any precautionary specification clause realize that these same electrical field forces can also be affecting the people who sit close to or on top of such powerful routeways as they 'swim' through the building structure. This is especially true of modern office buildings where the increase in telecommunications systems has dictated that modern office buildings are built like layer cakes with service floors increasing in depth and complexity.

Table 4.6 lists the field strengths of some common devices used in the home.

Table 4.6 EMF levels

Source of electromagnetic field	Distance from source (mm)	Field strength (mG)
Fluorescent lights, 10 W tube	300	1
Electric clocks	300	5–10
Hair dryers (according to time of use)	300	5–10
Electric baseboard heaters	180	23
Microwave ovens (information yet to be made available; in the latest models a leakage of 1 mW/cm^2 is permitted; caution is advised on siting)		

Outside the home personal radio transmitters and cellular telephones are becoming an attachment as important as a handbag or a hat. Quite likely, somebody may well be trying to design the new hat of the 21st century, complete with internal short-wave telephone and transmission set. These devices emit very high frequencies and are in the immediate vicinity of the human brain, itself a fine electromagnetic field force of the human kind. At the time of writing I have been unable to research any material on what effects cellular telephones can have on the human brain. My own common sense tells me that they may be more malignant than they seem.

Alternatives to the use of electricity

Modern-day governments are tending to equate the production of electrical power with a combination of political and fiscal power. In Britain a few years ago there was talk of building 20 nuclear power stations around our coast, otherwise all the lights would go out. This scenario has been dropped for a time. There is no question that electrical power is of great benefit. It can be produced safely and efficiently by the use of renewable energy resources, and transmitted by wire or cable, albeit extremely inefficiently at present. The question is, do we need so much power?

Posing this question we find ourselves in the realms of energy saving, a subject that is an important element of any matrix of designing for a healthy building, for it means that by a national policy of energy saving, the unsafe and unhealthy methods of generating energy power are reduced.

In general, while most new and existing UK office buildings — rarely air conditioned — use $0.5-2.00$ GJ/m^2 of delivered energy annually, new US office buildings in cool temperate climates, favouring low-energy use, now only use $0.3-0.7$ GJ/m^2, and several thousand truly energy conscious designs, as illustrated in Case study 3, Chapter 4, use from 0.05 to 0.20 GJ/m^2. All the US offices are designed to air-conditioned comfort standards, many without mechanical cooling equipment. The capital investment in these energy-efficient designs varies from negative to the equivalent of a $25-30\%$ return on investment.

Energy-efficient technologies extend from the rapid development of electronic ballasts for fluorescent and other high-intensity discharge lamps, to space-cooling techniques where electrical air-conditioners are not necessary, except in a small area of the country. The verdict seems to be careful solar control, energy-efficient lighting, coupled with good daylighting techniques, and, for the large-sized cooling plants, the use of seasonal ice storage.

It was the physicist and former nuclear weapons designer Ted Taylor who found a way to extract useful work from the water molecule. In the 1970s he began to use water to replace electricity. He took a garden hose and sprayed the water in a fine mist into the winter night air where it froze, not into ice but into slush, and then dropped it into an excavated pit in the ground, over which was placed an insulated cover. This ice slush was used later for his house cooling. Ted Taylor also found that when water freezes it pushes out all impurities and contaminants. This has proved to be correct. In 1990 on the northern shore of Long Island an iceberg was built, two square city blocks long and 100 ft (30 m) high. Its peak melting rate was in August, the point of maximum demand for electricity for cooling and fresh water.

Healthy Building Code Proposal 13
- Check if radon emission from ground or building materials is present
- Check electromagnetic fields from overhead or nearby power lines
- Check for underground water courses
- Wrap all electrical power points and light points with aluminium foil to reduce local emission levels of emf

Fig. 4.40   Swansea City 'plantasia'
(courtesy Rentokil Environmental Services
plc).

This technology for using the water molecule instead of electricity is now slowly being incorporated into new public and commercial buildings in the USA. An interesting example is the Prudential's 12,000 m² Enerplex South Building in Princeton, New Jersey.[28]

A late 1970s design used 0.31 GJ/m² of electricity per year compared with the 1979 US commercial building average of 0.5 GJ/m². It was then determined that this building could now be retrofitted with improved HVAC systems, saving 13% of energy; with improved lighting with electronic ballasts, saving another 13%; and with higher performance windows lopping off another 8% and reducing annual energy use to some 0.2 GJ/m².

The space cooling for this building uses seasonal ice storage as designed by Ted Taylor in conjunction with Princeton University's Centre for Energy and Environment Studies (CEES). Snow-making machines using compressed air plus 6 litres/sec of water can produce 20 tons of ice in winter which is then stored in an insulated pond of 12,000 m³. The system coefficient of performance (units of heat removed from the building per unit of electricity used — excluding HVAC fans) is about 18 compared with 3 for average refrigeration systems and 5 for the present state-of-the-art refrigeration systems.

The job of the designer, specifier and user is to seek ways and means of providing the energy and comfort we require without any ill effects either to ourselves or to our planet.

Visual appetizers and vegetation

The provision of plants in a building along with water has both a pleasing and healthy effect in that it naturally cleans the air of contaminants as well as providing simple cooling in conjunction with mechanical systems.

What is equally important is that you cannot just dump any old tree or plant in a room or internal light well and expect it to survive. Plants are like humans — they also require the correct balance of light, water and nourishment for healthy growth.

Many a beautiful interior can be ruined because the landscape designer and building services engineer assume that if the created environment is good for people then it is good for plants. The same misconception often occurs when the interaction is between machines and people. Some plants are known to be beneficial in the purification of hostile emissions from certain machines (desk printers and copiers); but the use of plants should not become a cheap way of dealing with the internal air quality.

The problem of knowing what questions to ask is the art of first learning what each team member knows of a given subject, and like every other specialist function described in this book, it is important that gardeners, landscape architects and other plant specialists are chosen with care. These people must also be prepared to understand the needs and requirements of architects, engineers and other people in the building design team.

The three areas of concern for the creation of a well-tempered environment for visual appetizers are:

* The condition of the soil and root surrounds.
* The condition of the air.
* The quality and amount of light.

Stephen Ashley[29] established some essential basic ground rules for keeping both clients and plants happy.

Light People operate quite efficiently at levels of light around 500–1000 lux. Trees, on the other hand, usually require 2000 lux minimum for 12 hours a day and may well prefer 50,000 lux or more.

There are two crucial factors, quantitative and qualitative, for light required for plants.

1 The first factor is the quantity of light needed for photosynthesis.
2 Second is the qualitative fact of night/day ratio and spectral analysis. Photosynthesis peaks at 450 nm (blue) and at 650 nm (orange to red). The blue light makes the plant grow in a particular direction while the red light makes it flourish. Too much infra-red stretches the plant.

Glass-covered atria have become part of the standard kit-of-parts of many buildings throughout the world. Yet it is interesting that if you visit the internal courtyards of Seville or Granada in Spain, which are influenced by Arab architecture, the internal space is open to the sky no matter what the external climate is like.

If the building services engineer, wishing to reduce the solar load, asks the architect for tinted glass in the overhead glazing of the atria, light levels will not be cut down, but the spectral quality of the light will be changed, which can affect the ability of plants to function. However, there are many plants that can survive in poor light. *Dracaena fragrans* can survive with light levels of 200 lux, and bamboo can live with light levels below 300 lux, similarly to their natural way of living in damp tropical forest undergrowths.

Another important aspect is to remember that plants are used to receiving sunlight at relatively low angles, in the area well above and below the equator: another item where the engineer has another item to think from a totally different perspective. Thus angled glass in atrias can help: in addition to having a dramatic effect on reflected light, it allows plants to grow well at different levels of a building. If artificial light is used it is important to remember that plants do not see light as we do. They react to different colours in different ways. It is vital to choose the right lamp. Incandescent lights are not good. They produce too much energy at the red end of the spectrum, which can simply cook the plant. Equally, the use of sodium lamps has to be done with extreme care.

Plants seem to like fluorescent lamps, especially the triphosphor varieties. Also, metal halide lamps have good 'vibes' for plants. Positioning of lamps must also be undertaken with care.

Plants and trees also react to heat, humidity and air movement in the same way as we do. It is important that at an early stage in any design process the kind of interior landscaping must be known, otherwise you could be looking for plants and trees, if you can find them, to suit your own internal climate. That is why step-by-step design actions, which are fully integrated and constantly related to form a 'synergy' of design actions are essential. This has been a powerful factor in the pleasing result of the NMB Bank, Amsterdam (see p. 95).

Plants can be found to suit most environments. In general, trees and plants are happy with temperatures between 13°C and 25°C. Humidity is not a problem, provided that it is also sufficient for human habitation. Gentle air movement is acceptable to plants as it is to people, and like us they don't like cold draughts.

Fig. 4.41 Arrayanes courtyard, Alhambra, Granada, Spain — 'the internal space is open to the sky' (photograph by Bill Holdsworth).

Fig. 4.42 Typical plants and trees within an atrium — Croydon (courtesy Rentokil Environmental Services plc).

The fact that plants reduce carbon dioxide and increase oxygen is a good thing, but sometimes large amounts of water breathed out can have a detrimental effect on the calculations made by the building services engineer. NASA have undertaken some work on the effects that density of species could have on building physics calculations. The publications *Landscape Industries International* and *Guide to Interior Landscaping* by the British Association of Landscape Industries can be of help.

See also visual appetizers in Chapters 5 and 8.

5 The building as a third skin: with international case studies

May it be delightful, my house;
From my head to my feet, may it be
  delightful;
Where I lie, all above me,
All around me, may it be delightful.

May it be delightful, my fire;
May it be delightful for my children;
May all be well;
May it be delightful with my food and
  theirs;
May all my possessions be well, and may
  they be made to increase.

Navajo*

*Source Mindeloff Cosmos *Navajo Houses*. 17th
Annual Report of the Bureau of American
Ethnology (1895−6) Washington GPO.

Like people, all forms of habitation grow and have flexibility, especially when they are constructed of 'living' materials. However, flexibility is often a problem for designers, constructors and the people who come to live in the built space.

The cities and towns and habitations we build fail in form, shape, usability, materials of construction and efficiency of operation and control. We have not thought of our work in designing and building as holistic, a comparatively new adjective describing the philosophical belief that the fundamental principle of the cosmos is the creation of wholes, i.e. self-contained systems of lesser or greater complexity. It is, in fact, a therapeutic system of design. In human terms it means that the totally human organism which is functioning as it should be either does not become ill or, if it does suffer sickness, makes a rapid spontaneous recovery, and that the occurrence of any illness, even one which seems confined to a particular organ, indicates that there is something wrong with the functioning of the whole.

International case studies

Case studies are presented from eight countries as examples of the changing state of the art in the development of buildings which are healthy and friendly both to people and to the environment that surrounds us.

Wherever possible the current up-to-date situation of those buildings actually built and occupied has been recorded.

Case study 1   Headquarters building, HLH International, Johannesburg, South Africa
Case study 2   Low-energy housing, Housing Work Yard Co-operative, Maaspoort, Den Bosch, The Netherlands
Case study 3   Night-flush cooling, Emerald People's Utility District Headquarters, Eugene, Oregon, USA
Case study 4   Housing Co-operative, Baggensgade, Copenhagen, Denmark
Case study 5   A user-healthy day nursery, Skarpaby, Stockholm, Sweden
Case study 6   Office building, project from Denmark
Case study 7   The NMB Bank Headquarters Building, Amsterdam, The Netherlands
Case study 8   Small country house, Great Wratting, Suffolk, England

Case study 1

Proposals for solar, total and regenerative architectural—engineering solutions for headquarters building: HLH International, Johannesburg, South Africa, 1980

Building architects: Louis Karol Inc.
                       Johannesburg, SA

ECHOES designers: Tielman Nicolopolis,
                    Architectual and
                    Solar Designer.
                    Bill Holdsworth,
                    Building Services and
                    Energy Engineer

26 10 South   28 02 East

The building was to be located on an island site in the centre of Johannesburg. The surrounding roads have heavy traffic creating noise and dust. The ECHOES designers recommended an immediate traffic survey to establish noise criteria patterns at different times of the day, which in turn would create design constraints on building detailing and engineering attentuation, as well as the location of working area.

The proposals were based on initial concept drawings of the architect, and due to lack of information many assumptions had to be made and proposals given, as to internal usage, forms of construction and other aesthetic considerations. The hope was expressed that the ECHOES designers were not trespassing on

Fig. 5.1   Perspective illustration of atrium showing water and plants filtering the air.

Fig. 5.2 'Surrounding roads have heavy traffic creating noise and dust.' Plan of HLH International building.

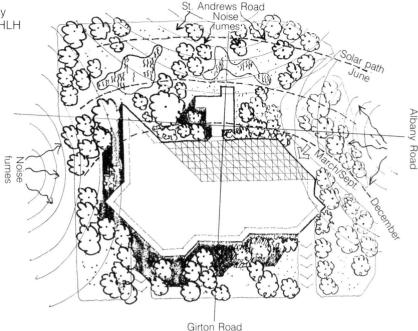

the architect's domain, but that our contribution increased the design options for the client, the architect and the eventual user.

To minimize the noise and dust, heavy tree and shrub planting on all sides of the building was recommended. Within such landscaping there would be night lighting for security, as well as fountains and paths of cooling water.

A densely planted landscape absorbs direct solar radiation, minimizes solar reflectance, and naturally filters, humidifies and cools, albeit in a rough form, the air passing through to be introduced into the building's ventilation systems.

The building lent itself to being treated as five separate zones:

1 Atrium
2 Rooms adjoining the atrium
3 Internal open landscape space
4 Outer perimeter
5 Rooms in the outer perimeter buffer space

Natural conditioners

1 Solar moats and atrium pools. Water to move along solar moats and channels all around perimeter of building and to cascade into atrium pool. Water to be introduced into moat by way of fountains connected to roof storage tank and to flow by gravity. Water pump to storage tank able to regulate water flow.
2 Vegetation. To be dense both externally and within atrium to assist cooling by means of shading and evapo-transpiration.
3 Solar collectors. Active system on roof to provide all hot-water requirements and for partial heating in winter.

Mechanical conditioners

4 Air-conditioning zones. Building divided into five zones. Each zone to have separate air-treatment method.

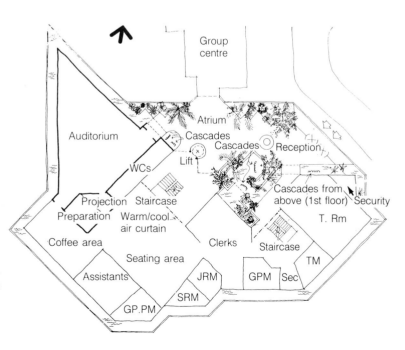

Fig. 5.3 Floor plan.

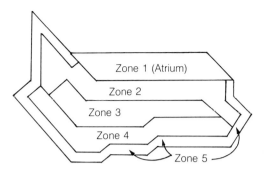

Fig. 5.4 The five zones of climate control.

Zone 1: The atrium   The atrium, because of its orientation, without effective control by the use of shading devices would have been extremely uncomfortable. The spaciousness of the atrium allowed the external landscape to flow into the building. In addition, cascading waterfalls from the first and second floors flowing into ground-level pools, together with large sub-tropical trees and shrubs would dramatically enhance the space as well as contributing to natural conditioning. This tempered air and roughly filtered air can then be introduced into either the mechanical ventilation system, or directly into open landscaped office areas with added conditioning at point of entry of cool/warm curtains.

During the summer air in the atrium heats up and rises naturally; it could then be rejected to the outside air by remote controlled ventilators in the roof, or recirculated into the building by mechanical means or natural stack effect.

If further cooling was required, mechanically cooled air could be brought in at low level over people's faces and bodies by localized cooling terminals in minimal quantities distributed by a 'puṇkah' effect.

During the winter outside air may well have been too cool for direct introduction into the atrium and this would be brought via rooftop mechanical ventilation plant.

Zone 2: Rooms adjoining the atrium   Because of the need of greater air change and occupancy in a proposed auditorium an individual mechanical full air-conditioning system was proposed, as well as heavier construction to stop traffic noise penetrating into the space.

The other rooms within the zone had a common factor in that their 'window side' was on the open atrium space. The rooms required privacy, and complete individual comfort control. Ventilated windows were suggested both for sound and temperature control. The windows could have been opened if required by the occupant.

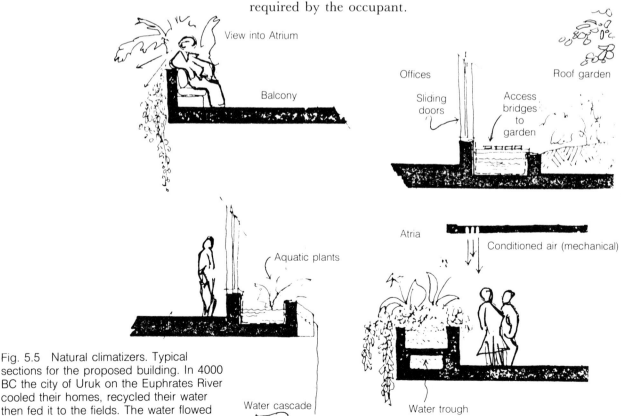

Fig. 5.5 Natural climatizers. Typical sections for the proposed building. In 4000 BC the city of Uruk on the Euphrates River cooled their homes, recycled their water then fed it to the fields. The water flowed along solar moats.

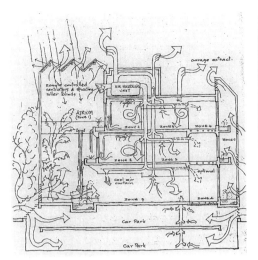

Fig. 5.6 Proposed AC system — summer operation. HLH Headquarters, typical solution.

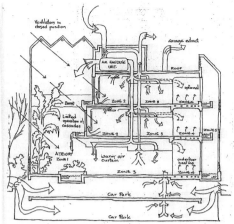

Fig. 5.7 Proposed AC system — winter operation. HLH Headquarters, typical solution.

Zone 3: International landscape office space   Air is allowed to flow freely through a central curve of open office space linked by stairs, bridges and balconies to the atrium, but able to be screened off by cool/warm air curtains at the atrium boundary, return air from this sector being returned to a buffer store.

Zone 4: Outer perimeter rooms   Private offices on all floors ventilated naturally through openable windows facing into the perimeter buffer zone. Natural air currents are cooled and humidified by the effects of slow-moving water, miniature fountains and planting.

Heating in the winter would have been either by a combination of 'dribble-heat' within the floor structure, or heating through convector grilles with the potential of added plenum air from the central ventilation plant.

Note: 'Dribble-heat' is a system of embedded tubes within a floor screed or upstand which maintains fabric warmth by a simple low-velocity flow.

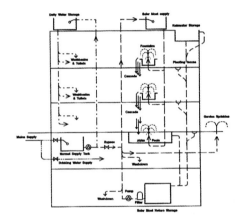

Fig. 5.8 Water recycling systems.

Zone 5: Perimeter buffer zone: This zone would incorporate a water channel or solar moat running along the perimeter of the building. Slow-moving water would be allowed to absorb direct and indirect solar radiation and by evaporation cool the air moving into the rooms.

The whole approach to building services was to see the building as an energy collector in the arrangement of space shape and construction to create an input of passive control.

The roof areas were seen as an important source of heat energy gain, which would be collected in a combination of solar ponds, solar panels (which in themselves also give roof shading) and solar sandwiches. As part of a policy of energy and resources recycling a water conservation scheme was proposed.

Natural daylighting was considered as a design objective, with the introduction of task lights.

Comment   This scheme, had it been undertaken, through many trials and errors of the design and build process, might have become a touchstone for many other designers, world wide, and an early example of a healthy building.

Healthy Building Rating * * * *
A design primer that 'bit the dust'.

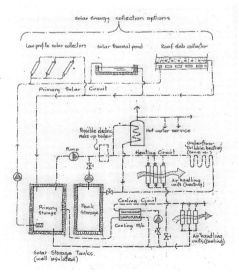

Fig. 5.9 Solar heating and cooling system.

Case study 2

'Letting the grass grow over your head.'

Low-energy holistic housing, for the Housing Work Yard Co-operative, Maaspoort, Den Bosch, The Netherlands, 1988

Architect: Archi Service, Den Bosch
Contractor: Biologische-Bouw Collectief,
              St Michielsgestel
Builder: Huijsmans Bouwtechnisch,
              Rosmalen

51 42 North   05 17 East

Architect Renz Pijnenborgh is a man who takes seriously the statement that a building must become your third skin.

The first protection is your real skin; second are your clothes; and third the place where you live and work. None of these elements should be a handicap for the human being. Real good clothes are made of pure wool, silk and cotton. All these materials have the same breathing properties as skin. This must be the same for the materials we use to build our houses and other places of human habitation. Renz Pijnenborgh

The first project undertaken as a base for building biology was a social housing project in the district of Maaspoort in Den Bosch, a city that seems to go in for innovation when it comes to house styles. In direct contrast to the strange space-age concrete balls inspired by an artist Dries Kreykamp and other developer-style houses are the grass-roofed timber houses of area MW2.

Fig. 5.10 'Space age' concrete ball houses inspired by Dutch artist, Dries Kreykamp (photograph by Bill Holdsworth).

Fig. 5.11 Low-energy housing, Maaspoort, Den Bosch, The Netherlands. Note the grass roof (photograph by Bill Holdsworth).

Fig. 5.12 Developer houses in Maaspoort, Den Bosch, The Netherlands (photograph by Bill Holdsworth).

Like most modern Dutch constructions, standing as they do upon reclaimed land, the houses have timber pile footings reaching to a stable sub-stratum.

The floors are conventionally framed — beams and joists support tongued-and-grooved boards. The underfloor space is enclosed around the house perimeter with timber siding, and the ground is covered with a membrane of PVC. The entire space is filled halfway to the joists with expanded clay pellets (perlite), providing insulation against ground heat loss. The remaining space to the tops of the joists is packed with clay loam, which provides protection from radon gas penetration. Details of walls and floor are shown in the illustration, which also shows the unique feature of an earth and grass roof. Supported on a strong timber frame, it consists of structural full timber boarding

Winter day (warmth from the solar-sun)

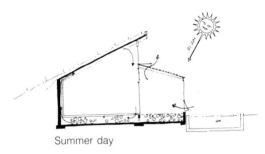

Winter night ot cold day

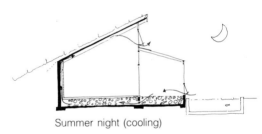

Summer day

Summer night (cooling)

Fig. 5.13 Earth housing and solar passive heating and cooling.

carrying a heavy gauge PVC membrane. Over this is a 100 mm layer of mineral wool, covered in soil. The grass is of a special variety that grows thick and even on a 45° pitch.

Grass roofs are typical in Scandinavia and Germany. They protect the roof from ultraviolet radiation and electromagnetic field forces. Also, the roof has a high thermal insulation value.

The use of loam allows the structure to breathe naturally, even though the walls are of a higher mass than with normal building methods. Cork is another material that serves as a natural breathing vapour barrier. Most of the internal walls, floors and ceilings are of wood, treated with beeswax. Finishing materials are linseed oil, natural resins, and paints that contain no harmful chemicals. Outgassing from construction materials is nil.

All electrical switch and lamp points are protected with a fine aluminium mantle to reduce electro-field forces. Heating is provided by a tile-stove with small-load skirting electric heating elements for topping up. The heating stove has a slow-combustion fire-box housed in a mass structure with high thermal inertia.

Comment   The houses in Maaspoort, Den Bosch, are delightful places in which to live and work. The community that lives in them are all self-supporting people. The benefits have been freedom from maintenance; low energy costs; minimal temperature changes in the roof surface; improved internal climate, with reduced diurnal change; improved sound insulation; filtration of dust particles from the air.

At first there was some scoffing from contemporary building contractors. The success of the venture has resulted in a major scheme of 'environmentally conscious' housing with a 100-unit scheme in Haarlem and the demonstration village of Ekolonia.

Healthy Buildings Rating ****
A primer for the future. An excellent example of a healthy place to live.

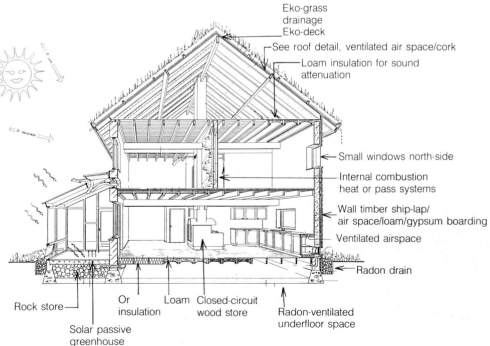

Eko-grass drainage
Eko-deck
See roof detail, ventilated air space/cork
Loam insulation for sound attenuation
Small windows north-side
Internal combustion heat or pass systems
Wall timber ship-lap/ air space/loam/gypsum boarding
Ventilated airspace
Radon drain
Rock store
Or insulation
Loam
Closed-circuit wood store
Radon-ventilated underfloor space
Solar passive greenhouse

Fig. 5.14 Section through biological house, Maaspoort.

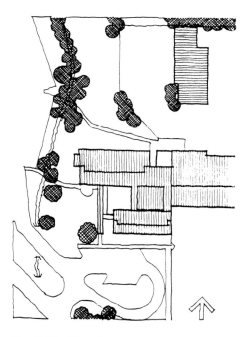

Fig. 5.15 Site plan.

Case study 3

Night-flush cooling for Emerald People's Utility District Headquarters, Eugene, Oregon, USA, 1988

Architects: Equinox Design Inc.
               John S. Reynolds;
               W.E. Group Architects and Planners;
               Richard Williams;
               Robert Caruthers

Daylighting
consultant: Virginia Cartwright   (drawings from Virginia
               Cartwright)

44 0 North   123 8 West

Climate: cool, wet and overcast winters, clear, dry summers with high diurnal temperature variations

An office building in the USA Pacific Northwest, with a floor area of 24,000 ft² (2230 m²) was designed to utilize daylighting, passive solar heating, and summer night-flush cooling of thermal mass in order to achieve significant annual energy savings. Comparisons of provision for air quality were made between typical contemporary office buildings and the one designed.

Early in the design process, daylighting was identified as an important design objective. This was because of popularity with the workers and its contribution to energy conservation by displacing the use of electric light. Daylighting provided the following design guidelines, which in turn produced the plan and cross section.

1 Almost all windows to face north or south, to avoid glare and summer heat gain problems of east and west windows.
2 Most work stations to be located within 2.5H of the windows, where H is the height of the top of the window above the floor.
3 Therefore the floor plan to be relatively narrow in the north—south dimension, about equal to 5H plus the width of a central corridor.
4 Ceilings to be high, with windows located up against the ceiling for deepest daylight penetration.
5 Light shelves to be used at both north and south windows.
6 More glass area to be provided above than below the lightshelf for penetration of more daylight and less glare beside the window with the result of a T-shape.
7 The window size was such that a 4% daylight factor was obtained.

Fig. 5.16 EPUD Headquarters (© Richard Alexander Cooke III).

Fig. 5.17   Building cross-section.

8 Individual daylighting controls, such as venetian blinds, to be available for the lower window areas, close to work stations.
9 Central clerestories provided increased daylight for work areas in the interior of the building.
10 Visual comfort was obtained by means of baffles, which ensured that no direct sunlight from the central clerestories reached any work station, and which also acted as acoustic devices.
11 Indirect lighting is provided for greater visual comfort, with the addition of individual task lights.

Energy conservation was provided by supplementary heating with passive solar energy leading to a saving of 27% of energy costs in a climate that is very similar to that of the UK.

Conservation of energy in the summer was achieved by flushing the building at night with cool air instead of operating compressive refrigeration by day to cool a building as is normally done in the USA (as well as in many other countries). Fans are operated at night to remove the stored daytime heat from the structure. It is important that adequate thermal mass must be provided in the construction of the building. For this building a ratio of 1.5:1 mass area/floor area was deemed sufficient. To increase the amount of night flushing the mass surface was increased by the introduction of hollow cores in the precast concrete slabs.

The large exposed surface areas of thermal mass in this predominantly open-plan office building could allow excessive sound reverberation. The vertical light baffles act to reduce this.

With daylighting being a key decision in the Pacific Northwest climate, the primary thermal task moves from summer cooling (largely because of electric lighting) to winter heating. To ensure that winter/summer levels of sun penetration are modified, deciduous vines on trellises are used to protect south-facing glass from the sun from early May to mid-October. The primary thermal advantage of deciduous vines is that they block sun at the autumn equinox (warm weather) and admit sun at the spring equinox (cool weather). Visually, the changing colour of the leaves (Virginia creeper, *Parthenocissus inserta* or *Parthenocissus quinquefolia*) affects the quality of the interior daylight in a subtle and pleasing way.

A typical designed-to-current-code Pacific Northwest office building might be expected to use about 158 $(kW/m^2)/year$.

Winter

Spring

Summer

Autumn

Fig. 5.18   Exposed surface areas of thermal mass (courtesy of Virginia Cartwright).

Fig. 5.19   T-shaped window.

However, this building was predicated to use 90 (kW/m^2)/year, but in the first year of operation used 126 (kW/m^2)/year, leading to more emphasis on ways to use the building by the occupants so that the full conservation potential can be reached.

(Information supplied by Virginia Cartwright and John S. Reynolds, Department of Architecture, University of Oregon, Eugene, Oregon, 97403, USA)

Comment   This is a building where the stress was laid on thermal, visual, and acoustic comfort, all taking precedence over indoor air quality. Smoking is banned in the main work areas.

Outgassing from the concrete and other materials worried the architects considerably, so the building was flushed out for a week before occupancy, and no air problems have arisen (July 1990), probably due to so much air volume per occupant.

The comfort system is VAV (variable air volume) ventilation with supplemental radiant heaters below the windows. No reliable ductwork cleaning equipment has yet been found, to sneak through the spiral wound ducting. The flushing air that passes through the building is filtered but the air into the hollow slabs is not.

External noise is blocked from a motorway by an old railway embankment. Occupants have expressed satisfaction with the daylighting.

Healthy Building Rating * * * *
An energy-conscious building design that is healthy.

Fig. 5.20   Emerald PUD office building.
Construction arrangement.

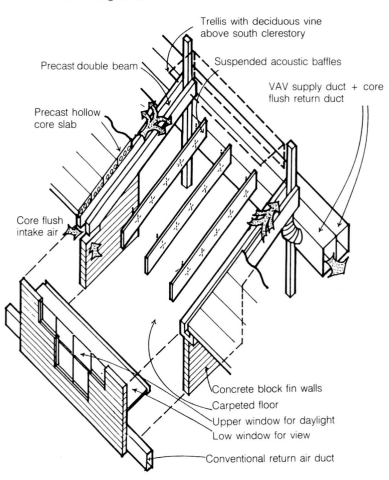

Fig. 5.21 A pleasant sunplace — a welcome and flexible addition (courtesy M. Kjærgaard).

Case study 4

Housing Co-operative, Baggesensgade, Copenhagen, Denmark

Architects: Floys Stein, School of Architecture, Royal Academy of Arts, Copenhagen, and Atelier ISYL

Energy consultant: Technological Institute, Tastrup, Denmark

55 41 North   12 34 East

Climate: An urban climate of a city experiencing coastal temperatures with mild winters and relatively cool summers. The micro-climate is affected by high buildings to east and north, a row of deciduous trees and a church tower to the west. These elements reduce the period during which building has direct solar exposure, yet also afford protection from cold winds, which increase in speed due to the tunnel effect caused by the physical obstructions (see solar radiation/temperature).

As part of a general urban renewal programme for a neighbourhood of Copenhagen, an apartment block, where the residents had already built a south-facing conservatory at ground level, was used as an extension of this solar passive experiment, completed in 1985.

The residents had already enjoyed the extension of their living space, in preference to the normal methods of installing new windows and increasing wall insulation.

The facade of any building is the result of many, often conflicting, requirements. In energy terms, it would have been desirable to have as much transparent area as possible, while the architect and client wanted to give the new facade a sense of its own individuality. The problem was how to improve comfort-of-living space with the use of 'friendly' materials but without the use of heavy machinery and while creating as little disturbance to the occupants as possible.

The new facade is supported on larch wood columns with floors constructed of:

1 40 mm magnesite screed
2 20 mm plywood
3 A cavity formed by 170 mm timber framework
4 10 mm cement/fibreboard for fireproofing

Ventilation air passes through the cavity and into the sunspace. Wind stability is achieved by attaching the decks to the corners of the building and the central stair tower. The new glass skin is 1.5 m in front of the existing building and creates a 10 m² sunspace for each flat. This sunspace saves energy in two ways:

1 By supplying direct and indirect solar gains to the heated spaces, thus reducing demand on normal space-heating
2 By providing a buffer zone with the external environment, partially to insulate against the winds, but also to provide a 'green space' within the apartments where plants and flowers can grow, giving natural filtration for any internal contaminants

The deck between the existing building and the new glass facade is placed 450 mm above the floor level of the flat. This allows as much direct sunlight as possible to enter the existing flat, and to provide the best view possible for people within the, now, inner room. The raised decks result in a great deal of daylight being reflected on to the ceilings, making it an added feature of passive solar design. Light-coloured floors also increase the amount of reflected light.

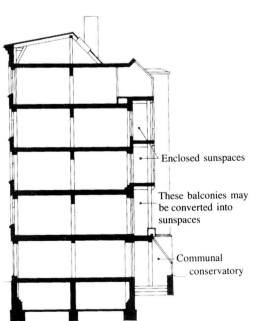

Enclosed sunspaces

These balconies may be converted into sunspaces

Communal conservatory

Fig. 5.22 On one side of the first floor, there is an open balcony rather than an enclosed sunspace

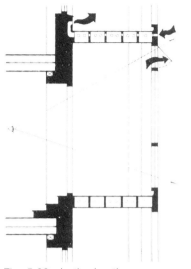

Fig. 5.23 In the heating season warmed air from the sunspace is convected into the flats. Excess warm air is vented to the outside in summer (source: Project Monitor 17, April 1988, European Community).

Comment   The architect, clients and occupants wanted single glazing for the new addition, since the inner windows are triple glazed. Such 'tightness' of living was found to be unacceptable. The building did not breathe enough. By allowing the new solar glazing to leak fresh air, any condensation on the inside of the glazed facade was reduced.

The user response has been high. Some tenants have installed blinds to reduce glare, others, automatic window-opening devices, for periods when they are away during the summer.

Since this project was completed (1985) the tenants have become more knowledgeable about how to use sunspaces. The sunspaces have reduced the energy load of the apartments by 34% and the internal air temperature is 5−10°C higher than ambient.

Denmark benefits from a policy of providing low-interest loans for refurbishment. On present costs (1990) the payback is 15 years. As energy costs rise, this will reduce.

Sunspaces not only offer energy savings, but can also contribute to buildings becoming healthy by:

1 Improving the living space, with the creation of a 'hanging garden'
2 Reducing condensation and mould growth, and giving tenants of city buildings a way of seeing that their health is connected with a building's health

Healthy Building Rating ∗ ∗
Although carried out in 1985, and primarily for energy conservation (Project Monitor No. 17, Commission of the European Communities), the project is a good example of retrofit work for inner city areas, that has brought some element of calm and belonging into ways of living in a city, with garden/sunspace that has protection from wind, rain, cold, excessive solar glare and noise.

Fig. 5.24 Construction of the external sunspaces (courtesy F. Stein Atelier ISLY).

Fig. 5.25 Completed facade (courtesy M. Kjærgaard).

Fig. 5.26 A user-healthy day nursery (courtesy Marie Hult).

Case study 5

A user-healthy day nursery, Skarpaby, Stockholm, Sweden

59 20 North   18 03 East

In the 1970s there was a big increase in the demand for day nursery places. The demand was met by fast-build methods. By 1980 complaints from staff and parents of ill-defined sickness led to some 30% of the nurseries exhibiting indoor climate problems. From the experience a user-healthy day nursery was constructed.

No one had foreseen the problems. No one was prepared. All the buildings had been designed to the appropriate Swedish standards (offered as examples of the state of the art in many other countries). There was no single clear cause for the problems that had arisen. Problems found were:

1   Gases and micro-organisms such as formaldehyde, 2-ethyl-hexanol and/or mildew were found in quantities great enough to constitute a health hazard
2   Moisture, and building materials that released easily volatilized pollutants
3   High indoor temperatures, resulting in low relative humidities
4   Air-tight structures, heat recovery and low air-change rates; inadequate HVAC systems

The day nursery for 15 children was built as a detached, single-storey building. The user area was 103 m^2 with a ceiling height of 2700 mm compared with the norm of 2500 mm.

The choice of systems and materials to minimize risk factors included:

1   Design of the foundation and other features to avoid risk of moisture penetration and/or water damage
2   Design of the ventilation system to incorporate heat exchangers that cannot transfer volatile pollutants from the exhaust air to the incoming air, and incorporating good air quality air filters, together with provision for allowing air flows to be increased above the required minimum values
3   Selection of homogeneous materials in preference to layered materials, and using screws and nails in preference to gluing, unless alternative 'healthy' materials are available
4   Avoidance of materials and designs which gather dust, or are difficult to clean
5   Avoidance of materials that can release pollutants such as hydrocarbons, formaldehyde and fibres into the indoor air (these are the materials that Dr Sherry Rogers describes as EI materials, i.e. those that create environmental illnesses: see Section 4.2)
6   Limitation of surplus heat from passive solar radiation by means of projecting roofs
7   Documentation of type and manufacturer of materials, paints, glues, mastics, etc. used in the building

Examples of practical solutions are given below:

Design criteria for a user-healthy project were:

To build a day nursery in which the design, the use of structural materials, space heating, ventilation, controls and lighting all interact to produce a good indoor climate for all seasons. The building services must be simple, reliable, inexpensive to maintain and energy efficient.

Exhaust-air ventilated windows   When air is extracted through the air space between the inner and outer double glazing of the window, no radiator is needed under the window. (Note: This concept was proposed by the writer for an office complex (Crown Estates) Drummond Gate, London, designed and erected in 1978–1981. Experience showed that the incorporation of a radiator skirting element would have helped the problem of cold air movement across the floor.)

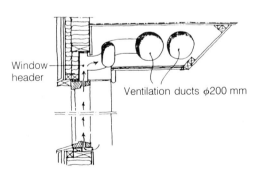

Window header

Ventilation ducts φ200 mm

Fig. 5.27 Exhaust air ventilated windows. In the airborne heating mode, the exhaust air is extracted through the space between the inner and outer (double) glazing of the window. No radiator is needed under the window (manufacturer's statement).

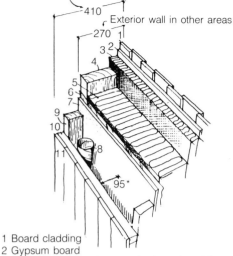

Exterior wall in experimental section

Exterior wall in other areas

1 Board cladding
2 Gypsum board
3 Insulation, 45 mm between horizontal studding
   Studding: 45 × 48 mm c/c 600 mm
4 Studding: 48 ×145 mm c/c 600 mm
5 Insulation, 145 mm
6 Plastic film vapour barrier
7 Gypsum board, 9 mm
8 Ventilating duct, ⌀80 mm c/c 600 mm
9 Studding: 45 × 95 mm c/c 600 mm
10 Nailing batten, 34 × 45 mm
11 Grooved and beaded board, 25 × 95 mm

Fig. 5.28 The exterior wall. In the waterborne heating with balanced ventilation mode, the air is extracted along the lower parts of the outside walls through spiro ducts in the wall construction.

The exterior wall   If a 'wet' radiator system is used with a balanced ventilation (space losses) mode, the air is extracted along the lower parts of the outside walls through spiral ducting in the wall construction.

Ventilation and heating   The experimental day nursery combined a flexible heating and ventilating system to facilitate various combinations of operating modes and setting.

The general rooms and rest rooms at both ends of the building were heated by either air or a low-temperature 'wet' radiator system, complemented in this experimental state by the addition of electric heating in the floor along exterior walls.

Thus three different arrangements were tested:

1 Radiator heating (fabric losses) with balanced variable-volume air flow from a corridor high-level discharge and thence extracted via a ventilated window
2 Radiator heating with fixed air-flow rate, and exhaust via the ventilated windows
3 Radiator heating plus outdoor air inlets behind the radiators, and variable air flow (traditional 1930)

Surface materials used were:

1 Ceilings
   (a) Gypsum panels (factory painted), 23%
2 Walls
   (a) Wood panelling, varnished with water-based clear varnish, 26%
   (b) Gypsum wall panels painted with water-based paint, 10%
   (c) High-pressure laminate on walls of wet areas, 7%
   (d) Wallpaper, 7%
   (e) Cupboards or directly-laminated chipboard, 4%
3 Floors
   (a) Linoleum floor covering, 15%
   (b) Plastic floor covering, 8%

Water-soluble paints, glues and fillers were used as far as possible, with solvent-based products used when no substitute could be found.

The main differences in the choice of materials in relation to those used in the standard day nurseries were:

1 Stained wood panels on walls in the children's areas and corridors instead of painted glass-fibre fabric
2 High-pressure laminates in wet rooms instead of wall-grade plastic covering or woven glass-fibre with extra coats of paint for water resistance
3 Grade E1 flooring chipboard (maximum 0.01% free formaldehyde) under linoleum. Grade E1 was also wanted for cupboards, but the cost for this project was considered too high and 0.04% free formaldehyde was used instead
4 Door handles of matt brushed brass instead of chromium plated

The costs were not so much higher than the norm. The small increase in costs led to reduced operating and maintenance costs, as well as to benefit in running and energy costs.

An assessment of the operating performance of the building for 1987 and 1990 is given in Table 5.1.

Table 5.1 Assessment of operating performance of the building

	1987	1990
Design temperature inside +20°C outside −20°C	OK	OK
Floor surface temperature +20°C	OK	OK
Temperature difference between head and feet < 3°C	OK	OK
Radiant temperature from windows etc < 10°C	OK	OK
Air velocity and mean air velocity under 3 min not to exceed 0.20 and 0.15 m/sec	0.02 OK	At some points higher than 0.015 m/sec
At 5 air changes, adjusted air diffusers		
Radon measured: low values	< 5 Bq/m^3	No other measurement
Formaldehyde: low values	0.01–0.03 ppm	Same as 1987
Reverberation time: frequency 500–2000 Hz	0.3–0.5 sec	0.3–0.45 sec
Chemical properties of air	All levels low	No change
Ventilated foundation	Some water seepage, plastic membrane tighter	Foundation OK
Staff health	Good	Good

Conclusions: 1987–1989 inspection.

Heating and ventilation The staff estimated that air quality became acceptable and pleasant when the air change rate was above five per hour. One exception was in the autumn of 1989 when the system was adjusted by lowering the air change to three per hour. Staff and children had expanded to 15 children and 4 adults per ward. There were also some faults in the ventilation system. The important lesson is that air quality must relate to changing occupation and use.

Complaints of draughts in the window zone when air changes were higher were noted by staff. It is believed that suppliers of low-velocity air equipment are rare. The all-air heating system with five air changes with varying degrees of recirculated air (0–40%) was considered by the staff to be acceptable. When the recirculated air was higher than 60% then the staff considered that the air quality was unacceptable.

Other problems were found with an all-air system. The final installation is now a combination of hot water radiator system and balanced ventilation system.

Choice of material The choice of material gives a low volatile organic level. In the first year of operation some particles of 'acrylate' were found in the linoleum flooring and some of the water-based colours. This pollution had, however, disappeared after 3 years. In the third year some outgassing substances were observed from timber boarding, but measurements were very low. The last measurement to check VOC emissions was undertaken in March 1990 when it was found that the total amount of VOC in the large playrooms was 0.09 mg/m^3 and in the small playrooms 0.11 mg/m^3. The external concentration was 0.05 mg/m^3. A measurement was also undertaken at night when the ventilation system was turned off. The percentage VOC then rose to 0.28 mg/m^3. All measurements were undertaken by the Swedish Government's Testing Laboratory.

Construction The investigation revealed that further study is required by manufacturers and designers to produce a suitable floor construction for a day nursery: one that can promise security against moisture and mould, and yet still provide a warm floor.

The light fibre construction has on the other hand given complete satisfaction for the indoor climate, both for winter and summer. There has, however, been a call for sunblinds on the sun-facing aspects to eliminate the summer heat and sun glare.

Marie Hult, Stockholm, 1 August 1990

Comment As an experiment in healthy building design this day nursery has shown that by the use of non-polluting and homogeneous building materials, coupled with good air movement without draught, and without cold floors (adding the radiant perimeter strip was important), a simple traditional HVAC system can do the job.

A full-scale investigation project, undertaken in real-life conditions, created a viable discussion climate between all parties involved in the provision of day nurseries, as well as being beneficial for other sensitive uses. Some of the solutions have already been employed in other building programmes. This applies to demands for a higher air change rate in children's playrooms, raised foundations, higher ceiling heights, and different types of heat exchanger.

Healthy Building Rating * * * * *
An essential model for the development of any healthy building code and materials specification.

Case study 5 is based on work undertaken by J.V. Andersson, Scandiaconsult AB, Stockhold, 1988, and Marie Hult, Architect, Stockholm Social Welfare Administration, 1990.

Case study 6

Integrated utilization of passive solar energy and natural daylighting in an office

Project for a 4400 m² three-storey office building which was awarded First Prize in the May 1989 CEC Architectural Competition: 'Working in the City'

Architects: KHR a/s Copenhagen, Denmark
Contractors: ISLEF, Denmark
Consulting engineers: CENERGIA Aps and Esbensen, Copenhagen

55 41 North   12 34 East

Commercial buildings are traditionally dependent on artificial lighting and large mechanical ventilation and cooling systems. Glazing areas in the facade are a key factor in the process of minimizing the total energy consumption of the building.

The diagram illustrates two significant features:

1 The optimum glazing area for heating or cooling or daylighting differs significantly.
2 The benefits from utilization of natural daylight have a very strong impact on the optimum window area, when the total fossil energy use of the building is considered.

As with the Emerald People's Utility Building, in Oregon, USA (Case study 3) the right mix of passive solar energy, daylighting and natural ventilation strategies can only be achieved through an integrated design approach.

Fig. 5.29 Model of award-winning healthy building by Poul Erik Kristensen, Planum International Ltd.

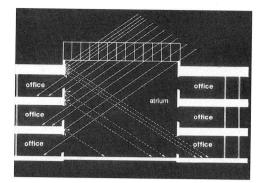

Fig. 5.30 The smaller window area at the upper level, 40% of wall area, leaves a larger opaque wall area, which reflects daylight into the atrium (courtesy Paul Erik Kristensen).

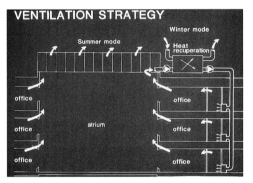

Fig. 5.31 Ventilation strategy (courtesy Paul Erik Kristensen).

Architectural constraints resulted in an L-shaped building with a curved entrance. To use the best of northern hemisphere winter and summer sun paths, the building was turned towards the south-west.

For this project compared with the NMB Bank Amsterdam (Case study 7) annual heating loads did not show any marked difference. The ability to use the solar aspect was limited, due to the 'boxed' shape of the building. However, the fact that windows are on the cooler side of the building does help where overheating problems from too many electric lights are easier to control.

Indoor temperatures were between 20 and 25°C, and night ventilation (a flushing technique) was very efficient in bringing down room temperatures.

The prime objective of the scheme was to reduce energy consumption; computer models predicted a 50% reduction.

The atrium acted as a buffer; although the heating load was higher than need be, the electrical load was much less, and the saving mainly to do with lighting.

Comment The design shows that low consumption of fossil fuel energy can be achieved in a building together with improved thermal and visual comfort. Natural daylight and natural ventilation are important factors in making buildings more energy conscious.

The building ideas are in themselves traditional. They seemed to be radical, because we had for a long time lost the art to conserve in building.

There was no evidence available that materials of construction had been seriously taken into account in respect of their energy content, or pollution factor.

Had the building design moved away from the concept award stage, then with the knowledge currently coming on stream, the architect and his team might have been even more adventurous.

Healthy Building Rating * *
Energy-conscious design with the right idea.

Case study 7

The NMB Bank Headquarters Building, Bijlmermeer, Amsterdam, The Netherlands, 1987

Client: NMB Bank Amsterdam
Property development/management: NMB/MBO NV, Amsterdam
Architect: Alberts and van Huut, Amsterdam
Construction engineers: Aronsohn BV Rotterdam
Building services engineers: Treffers and Partners, Baarn
Landscape consultants: Copijn BV Utrecht
Acoustic consultants: Peutz, The Hague
Interior designers: Billing Peters Ruff, Stuttgart
                        Theo Crosby, Pentagram, London
Main contractor: Voormolen-Heijmans-IBC, Amsterdam

52 23 North   04 54 East

To the south of old Amsterdam on reclaimed land from an ancient lake stands the headquarters of one of Europe's largest banks, the Nederlandse Middenstandsbank (NMB). Designed

Fig. 5.32 NMB Bank, Amsterdam, The Netherlands, final scheme.

Fig. 5.33 The main entrance from a pedestrian square of the NMB Bank (courtesy Alberts and Van Huut Architects).

Fig. 5.34 NMB Bank — internal light/stair well (courtesy Alberts and Van Huut Architects).

from the anthroposophical point of departure that states that nature provides the example to be followed by architecture, the result is in complete contradiction to the rectangular symmetry found not only in The Netherlands but also in every other look-alike world city.

Ten linked sloping towers stretch sinuously towards the sky, crowned with an azure blue pentagon. Brick walls slope and change direction, leaving few right angles.

Inside, the building opens into a glorious cavern, not measureless and certainly not sunless. Here is a spectacular street, twisting and turning, varied in every dimension, 350 m long, full of light reflected off soft coloured surfaces. People look relaxed: a world of difference from your narrow confined stuffy office corridor, with a multitude of doors opening on to uniform little boxes. Natural light spills on to the internal circulation street from the towers. Rainwater, collected, stored and filtered on its journey to feed the many internal plants, meanders down a wide flow-form balustrade at the side of the main staircases, slowly evaporating and cooling the air.

In each cluster block are the offices, where everyone has the right to space next to a window which opens. Good daylighting was provided by ensuring that the top section of the external windows had deflecting louvres for the daylight to be reflected into the inner area of offices, with external solar blinds and louvre drapes on the southern faces of the building. Specially designed light fittings give both ceiling-reflected light and downward light,

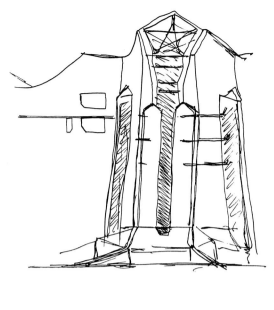

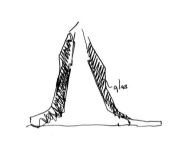

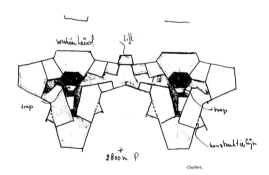

Fig. 5.35   Design sketches by Ton Alberts.

with electric lights close to the windows programmed to cut out when natural daylight is sufficient (about 500 lux).

The bank building is unique in many ways. The building has proved to have one of the lowest rates of energy consumption in the world. For a gross floor space (including internal car space) of 78,000 m², housing 2500 people who use 21 lifts, eat in four restaurants, and rely on the 'brain' of every modern bank, a computer centre with massive and flexible use, the total energy load is 111 (kW/m²)/hour, with a possible reduction to 98 (kW/m²)/hour. All for the total construction cost of £1000 per m².

Another departure from normal design practice was the method of selection for both site and architect, together with his team. It was a democratic decision by the bank's employees on where the building site should be, followed by a competition when every Dutch architect worth his salt was asked to submit ideas. All were vetted, and candiates on a short-list were then interviewed by a committee panel elected by the bank's staff. Office cleaners sat next to top accountants. Each had a say. The result was that before a pencil was sharpened, the different specialists were brought together. Engineers, structural and physical, acousticians, architects, landscapers and accountants, rubbed shoulders with contractors, artists and occupational therapists, all of whom worked and debated with the representatives of the people who were to work and live for a good part of their lives in the building.

A radical beginning for any design process. It was from this interaction, this synergy, that the serpentine shape of the building came about. The normal rectangular block gave way to the elements of wind, light, sun aspect, traffic noise and other city pollutants.

Each turn of shape brought about an optimum solution for work use, architectural sense and energy saving. Natural cooling, and passive solar energy aided by the active pentagon-shaped solar collectors at the tops of the towers, coupled with highly efficient natural-gas turbine generation of electricity, with the waste heat recycled and stored in large accumulators, was to result in an exceedingly healthy and low-energy-use building.

The building has a structural facade of a thermal sandwich: 180 mm brick, 30 mm air space, 100 mm insulation and 180 mm internal concrete facing. The infiltration rate is 0.1 litres/sec. A good characteristic of concrete (provided that it is radon free) is that it has a stabilized temperature for a long period. In the hot summers (30−35°C) the night temperature drops to + or −17°C. Cool air streams are allowed to flush through the building at night, cooling off the concrete. Should the building get too hot during the day air is cooled through the ventilation system, using the dissipated energy from the generation of electrical power.

Comment   Building contractors and installers tendered in the time-honoured style, but they were also party to the same humanistic interaction. No idea was ignored.

The NMB Bank project illustrates an organic approach to design both in respect of the building method, and in the democratic decision-making shared in by the bank's employees.

Although no deep study had been made on pollution from materials, the use of night-flushing of fresh air through the building not only helps energy planning, but also has the added advantage of expelling any contaminants.

According to the architect Ton Alberts, the staff are happy

Phase 1
Traditional block-form construction; less
than 50% of fenestration obtains sunlight from southern aspect. Wind impact
causes scrubbing and increases heat loss.
Solar radiation is higher, leading to
increased cooling loads. High-impact
sound from noise sources.
Estimated energy load: 280−300
(kWh/m²)/year

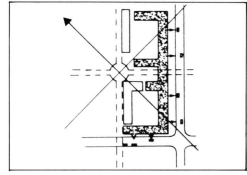

Phase 2
Some natural shaping added, with less
exposed surface on northern side.
Estimated energy load: 250−280
(kWh/m²)/year

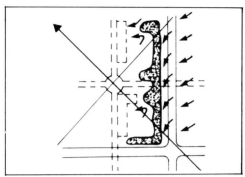

Phase 3
Break-up of traditional block form on
south-east and south side, increasing
area of fenestration for both sunlight,
and natural daylighting. Wind impact
broken up, less noise. Service space on
northern side.
Estimated energy load: 200 (kWh/m²)/
year

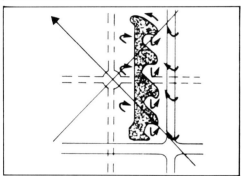

Phase 4
Building broken up into blocks to allow
wind to move through gaps. Greater
allowance of fenestration for natural
daylight. However, greater heat loss,
and solar gain with vertical surfaces. In
energy terms the advantages were
cancelled by disadvantages.
Estimated energy load: 200 (kWh/m²) /
year

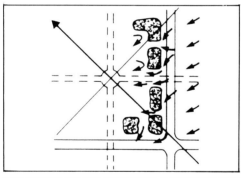

Phase 5
Final scheme consisted of ten cluster
blocks connected together. Vertical wall
surfaces were inclined to obtain maximum benefit of solar inclination; it
helped to reduce the impact of traffic
noise, and expressed the building's
essential earth-bound character.
    It also allowed each cluster block to
be set in the best relationship to all the
climatic considerations, and walking
distances were reduced by 50%
Estimated energy load: 98 (kWh/m²) /
hour]

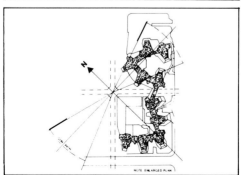

Fig. 5.36 Synergy study.

Design criteria:
Winter −10°C to +20°C with +3°C
swing
Summer +28°C to +24°C
Heating capacity: 3400 kW
Cooling capacity: 2400 kW

Fig. 5.37   NMB Bank — flow form design by English designer Peter Rawstone (courtesy Alberts and Van Huut Architects).

with the building, a view that the author can confirm from his own investigations. The NMB Bank has also declared that since the occupation of the building staff sickness has decreased by nearly a quarter.

Special investigation was undertaken to check if the material used for the floor, Tranie, is resistant to non-polluting cleaning materials. It turned out that the floor can be cleaned year after year; neither the surface nor the colour of the Tranie floor will change.

Indoor air quality is very good, even though the main bank area has no mechanical cooling. The good indoor air quality is regulated by a combination of flow-forms and internal planting. A thorough investigation has taken place to see how many leaf-green plants were needed to give the air its humidity and proper oxygen percentage content.

Healthy Building Rating * * * * * * * * *
The NMB Bank is a fine example of a clean healthy building which has nothing but a positive influence on the well-being of the people working there, and the people who live and work in the surrounding streets and squares.

Case study 8
Small country house, Great Wratting, Suffolk, England, 1990

Architects: Busch, Masheder Associates, Saffron Walden, Essex, UK
Builder: Carr and Bircher, Saffron Walden, Essex, UK

52 04 North   01 09 East

Fig. 5.38   Country house, Great Wratting, Suffolk, England (courtesy Hartwin Busch).

Hartwin Busch, an architect, had studied in Germany and moved to England in 1979. While building his own house he became very ill for three months after having applied several wood finishes to a floor. He had made no connection between his illness and the floor varnish, when again he became ill while using the same material for kitchen tops. It was only when a friend visiting from Germany learnt of the product Busch was using that the connection was made.

Reference is made in Section 4.2 to the work of the Burger Hospital, Stuttgart, and to the work of Dr Sherry Rogers in the USA, where there is definitive proof of high-toxity emissions from many sealants and varnishes that are on sale.

Learning that he had been taking years off his life, Hartwin Busch changed the direction of his architectural practice to thinking in terms of building biology.

The traditional English country-style house was conceived as an adaptation of the principles of building biology to the situation and building practice found in England.

Building materials were researched and selected for the way in which they affected the user's health and general well-being. Where possible preference was given to locally available materials; the single exception was paints, stains and surface finishes. The German import product Auro was used (see p. 131).

The house has been built in a rural area, and the overall design reflects traditional building patterns, with a steep roof pitch of 54° and shallow depth of building of 4.8 m, this being the maximum economic span of timber, with brick and rendered outside walls.

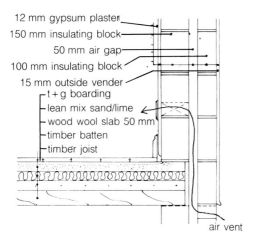

12 mm gypsum plaster
150 mm insulating block
50 mm air gap
100 mm insulating block
15 mm outside vender
t + g boarding
lean mix sand/lime
wood wool slab 50 mm
timber batten
timber joist

air vent

Fig. 5.39   Construction of wall and ground floor (drawing by Hartwin Busch).

Choice of site   The site opens to a magnificent south-westerly view over a river valley. The house position is just below the crest of the hill, giving protection from harsh easterly winds.

Position on site   Before the house was staked out, the site was walked with dowsing rods to establish areas of geopathic stress (see Section 4.1). As a result the house was shifted a few metres downhill.

Construction   To avoid the use of an oversite concrete slab, the method found throughout England in the By-Law houses of the late 19th century of timber-boarded floors on timber joists spanning the outside wall was used. The added late 20th-century improvement was the inclusion of 50 mm woodwool insulation slabs with some lean mix on top. In this way the ground floor is well insulated with some thermal storage capacity and all the positive properties of wood: breathing, moisture balancing.

With the wall, emphasis was placed on a balance between good insulation, thermal storage, and diffusion, thus allowing the wall the ability to permit air renewal through its own fabric. The insulation block Durox was chosen for its low radioactive emission.

Finish is a two-coat pure gypsum plaster which has excellent diffusion and hydroscopic properties but posed a minor problem because the builders were not used to its relative softness.

The roof was of traditional timber construction with clay tiles.

Services   The boiler was mounted on the outside wall to avoid fumes and noise and to save space. The heating system put emphasis on radiation rather than convection to allow for comfort levels to be reached at lower air temperatures.

The provision of a 'demand switch' is an optional extra so that the occupants can avoid the generation of electrical fields at time when no electricity is being consumed.

Comment   The house benefits from not being too radical in introducing concepts of building biology to England, although such radical thinking is only so to recent generations who have not questioned the use of building materials or methods of construction.

In this area of eastern England, there are many examples of houses built as early as 1400 using similar materials and forms of construction. But these ways and skills were becoming lost. The house was built for sale on the current housing market.

The local builders Carr and Bircher deserve praise for bringing to the project skill, care and enthusiasm.

The principles of returning to local knowledge and the roots of knowing how to live in concord with nature show a basic common sense.

Healthy Building Rating * * * * *
A positive contribution and example of high-quality healthy housing that can be economically viable in a volatile housing market.

Pointers to the future

The international case studies presented are pointers to environmental and energy-conscious building design. Throughout the world there are many new developments taking place. There have also been some ideas that remain in the background, yet are worthy of being illuminated for all to see and act upon.

Pointer 1:   West Germany and Poland

Unlocking earth's energy   There are many examples, world-wide, of using the inherent latent heat-energy of the earth. I may be wrong, but it is my belief that the great maze of passages behind the solar passive wall at the great palace of Knossos on Crete, built 1000 years before the hanging gardens of Babylon, was a simple device for good heating and ventilation, with an overtone of religious significance to make it all the more magical.

Another bit of magic was undertaken in 1876 when a John Wilkenson patented a 'sub-earth ventilation system' to cool a dairy. Even earlier, in 1858, a Dr Jeffrey proposed an earth-cooling system to ventilate a soldiers' barracks in India.

We may view the uppermost 50 feet of the earth's surface as one vast equalising reservoir, ready to absorb a large proportion of summer heat. If we adopt proper measures for cooling thoroughly in the winter this mass of earth, we may have brought it down nearly to the winter mean before the ensuing hot season, ready to absorb again much more heat than when it had to cool itself by tardy spontaneous upward conduction through the whole mass.

Dr Jeffrey's plan required a plot of 120 × 80 m with some 200 wells dug all over at around 7 m intervals. There is no record of Dr Jeffrey's refrigerator-well ventilation having been a success.[1]

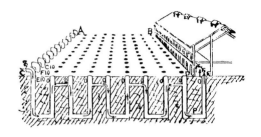

Fig. 5.40 Dr Jeffreys' earth cooling system (India, 1858). (Source: *Building Services Journal*, June 1989.)

Germany

Some years ago, Eugen Mayer, retired chief engineer of the town of Heilbronn, decided to make use of his cool cellar. He built vertical ducts up to his house which served as ventilation in the summer. Later he added a thermal store to reduce his winter heating costs.

In the 1970s, while engaged on ways to make energy savings in the town's schools, he observed that the summer, daytime air in the Elly-Heus-Knapp school entered at one end of a 60 m underground service corridor at 31.5°C and cooled down to 22°C.

Thus by allowing fresh air from the subterranean corridor to be integrated with the ventilation system, classrooms on the south and west side of the building could be cooled under the ruling limit of 26°C without any mechanical facilities. Similar efficiencies were noted for heating in the winter months.

Thermolabyrinth   During the planning in 1970 of a new theatre, the Heilbronn City authorities obtained grant aid from the Federal Ministry of Research and Development to investigate the utilization of the earth's heat both for summer cooling and winter warming.

The theatre finally constructed has a major auditorium for 700 people, a music room seating 150, extensive stage facilities, exhibition foyer, administration offices and underground garage.

Fig. 5.41 The U-shaped labyrinth with the heat exchange surfaces shown on the bottom of the duct. These 'teeth' were kept low in profile to limit drag coefficient. (Photograph by Bill Holdsworth.)

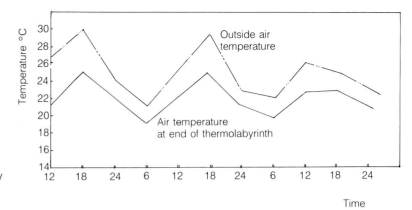

Fig. 5.42 Thermolabyrinth system. Energy balance for three consecutive operating days (19–21 August 1985).

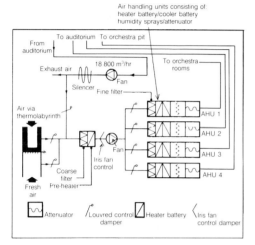

Fig. 5.43 Theatre Heilbronn. Schematic drawing showing the main air handling units for the system.

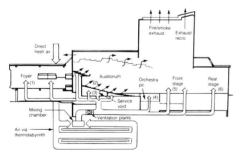

Fig. 5.44 Diagrammatic arrangement of the earth energy system, Stadttheater, Heilbronn (courtesy Mayer and Holdsworth, *Building Services Journal*, June 1989).

Twenty-five separate air-handling systems with a total load of 155,000 m³/h were computed. From transmission calculations and air rates it was determined to have on-site field trials. The cost of the grant allowed for the construction, in the foundation, of a concrete U-shaped labyrinth 6 m wide by 1.35 m high. The total canal surface was 1250 m². To increase the heat-exchange surface an addition of 450 m² was obtained by the introduction of 'dragon's teeth' on the bottom surface. These teeth were kept in profile to limit the drag coefficient, the air speed inside the labyrinth being between 2.5 and 4.0 m/sec. For transitional air humidification and cleaning, a piping system with spray nozzles was to be provided.

Six-and-a-half years after its installation I found the lack of dirt impressive. It was robustly built, but sensitive in operation; 72 measuring plates had been built in the concrete passages to record temperatures of the air and surfaces of the wall and floor, with the results being fed into a main comfort conditioning programme controller, which in turn regulated various control valves and motorized louvred gates.

All 25 individual ventilation systems are connected to both immediate outside air, and the air that is allowed to pass through the thermolabyrinth.

A detailed evaluation made by the Association of German Engineers Inland Climate Division (VDI 22078) proved that on a summer's day with a maximum temperature of 32°C some 2700 kW/h of the 3000 kW/h required, or 90% was provided by the air passing through the concrete labyrinth where it lost its heat into the building mass. A recording of the differential air temperatures for three days in August 1985 is a good visual illustration of this simple and inexpensive concept.

Stadttheater Heilbronn

Architect: Rudolf Biste + Kurt Gerling, Berlin
Building services engineers: Brandi Engineers, Gottingen
Consultant: Eugen Mayer

Source: 'Unlocking Earth's Energy' is an abstract from the author's article published in *Building Services Journal* (UK), June 1989.

Poland

Retro-fit earth energy use Leaving fresh air input to the variables of the weather had led Gerard Besler and Jaroslaw Dowbaj of Wroclaw Technical University Engineering Institute to seek solutions for a forced-air ventilation system that relied on the free renewable energy found in the ground.

Within the grounds of the Engineering Institute a full working

Fig. 5.45 Wroclaw: schematic showing the earth heat-cool labyrinth.

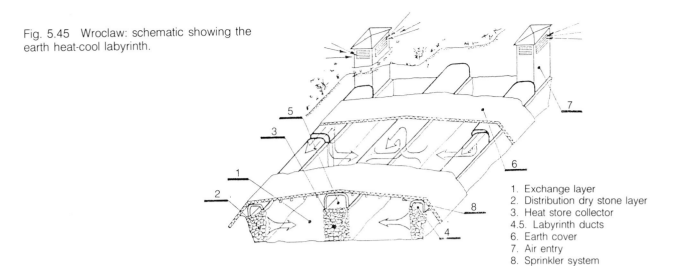

1. Exchange layer
2. Distribution dry stone layer
3. Heat store collector
4.5. Labyrinth ducts
6. Earth cover
7. Air entry
8. Sprinkler system

1. Fan
2. Rock-bed
3. Solar collector
4. Auxiliary heater
5. Air filter
6. Air in
7. Air out
8. Dampers

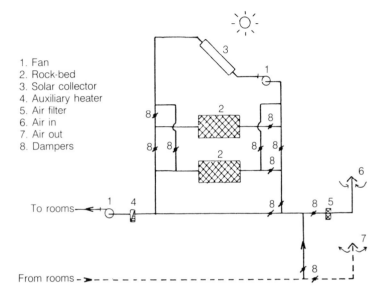

To rooms

From rooms

Fig. 5.46 Wroclaw. The solar air heating/cooling system.

model was constructed. A labyrinth was built which supplied external air to a filtered ventilation system. The addition of an active solar panel provided another source of free energy.

The use of solar energy in Poland meets with economic and technical problems because of the very little insolation (approximately 3600 (MJ/m^2) per year). There are about 550 solar hours from October to April, while the heating period has 5088 hours, giving only 10% of usable solar energy in winter. But this winter energy when coupled with the heat extracted from the thermal earth store leaves only a marginal top-up from the district heating system. The winter air temperature rise from the solar collection often exceeds 25°C, with maxima of 40°C.

The conceptual system resulted in outside air temperatures of −20°C being raised by the earth maze or labyrinth to 0°C, with a further rise of 10°C when the air was passed through the solar panel and then through a cellular concrete floor slab. Equally, the system was, like the example of the Heilbronn Theatre, also responsive in the summer when an average air temperature decrease was 10°C. From concept to practical application has led to the idea being incorporated in a test block of workers' flats in Wroclaw.

Results are still being collated. But the method lends itself to a whole series of beneficial ideas in both new and existing

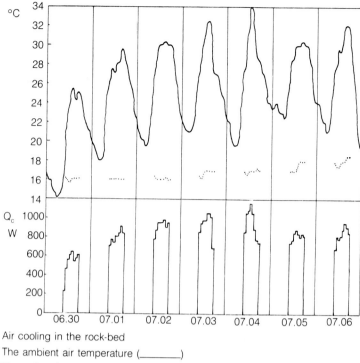

Fig. 5.47   Air cooling in the rock-bed.

Air cooling in the rock-bed

The ambient air temperature (_____)

The outlet air temperature (. . . .)

Q_c — cooling power of the exchanger

structures. Coupled to underground service grids it could lead to the introduction of a new element in town and city infrastructure planning: the building-in of earth systems as a major energy-saving component.

Scientific tests are now being assessed in Poland on improvement to air quality by the use of underground membranes. In the tests carried out on the Wroclaw University prototype the amount of bacterial impurities in the air passing through the rock-bed exchanger decreased from 3789 bacterium cells/m³ to 2000. Final results comparing the nature of various rock materials and the composition of the contaminants are not yet to hand; but the results so far are encouraging.

Source: Technical University Wroclaw, Poland. Incorporated in a paper given by the author to the RIBA/Department of Energy Symposium on the Architect, Energy and Global Responsibility, RIBA, London, January 1990.

Comment   Earth heat/cooling systems using simple earth-glazed pipe buried 1 m below the earth have been used in both Ohio, USA and Wales, UK for the conditioning of animal housing. I am sure that the system used by the British Army in India in 1858 worked, and is possibly still doing so.

Pointer 2: Sweden

Kristallen: the office building of the future, 1990   This was a seminar held in Uppsala, Sweden in May 1990. Dios Properties AB, sponsored by the Swedish Council for Building Research (Evaluation of daylighting and energy consumption), are building Sweden's first full-scale healthy building. All materials being used will be harmless to the environment and the future inhabitants.

The structural part of the building is concrete elements in the exterior facade and concrete beams and columns towards an inner facade. The structural part of the floor is hollow-core concrete floor slabs.

The facade is covered with bricks in various shapes, which gives the facade an attractive and varied expression.

As with so many buildings today, there is an atrium where the walls are covered mainly in maple plywood and wood wool elements. Floors in the circulation areas are covered in cut natural stones and ceramic tiles. In the office spaces linoleum is used. One of the main objectives with Kristallen has been to create a clean and healthy building.

The heating and ventilation system is simple, yet flexible. Individual control of heating and cooling in each office room will be possible. Special consideration has been taken to secure a system with a low noise level and no cold draught. The ventilation system will secure an air change rate of 1.5 litres/h. There is no air recirculation to achieve maximum air quality. All air ducts are constructed so that they can be regularly cleaned.

Client: Dios properties AB
Developer: Donald Ericsson AB, Uppsala
Architect: Contekton, Helsingborg
Consultants: Esbensen, Copenhagen

Comment   From the information obtained, it seems like a first step in making sure that developers are not worried by using ideas that have not been tried and tested.

During the author's apprenticeship with J. Jeffrey's and Co., London, 1945—1950, all air ducts had cleaning doors. In 1979 the design for the first phase of the Crown Estates Commissioners Drummond Gate building (completed in 1981) had individual modules that had independent heating and cooling, by the adaptation of the Svenska Flakt Combivant system. All the materials were 'human' friendly. At the time the clients and their project managers were difficult to convince. Smiles all round now!

Pointer 3: Denmark
Low-density affordable housing for allergy sufferers, Lejerbo, Denmark, 1990   The Danish Building Industry Development Board has subsidized the very careful planning of an estate of 11 low-density affordable houses for people suffering from a number of different allergies.

The object of the project was to examine the evidence for a possible reduction of environmental problems in modern tightly sealed dwellings, by the use of carefully selected building materials and increased ventilation.

Each dwelling has its own ventilating system, ensuring a complete hourly air change. The system is equipped with a pollen filter. In wet rooms, the exhaust function has been increased by 10%. A heat-recovery system keeps the energy requirement down to a level only slightly higher than in houses with less frequent air change. Interviews with tenants have indicated much satisfaction with the indoor environment.

Further information: Bent Olsen, Soren Olesen AS, Dalgas Avenue 46 DK 8000, Arhus C. Tel: 45 6 11 20 00

Comment   Such housing will be in greater demand as the cocktail effect increases: an area of health concern that has so far been ignored.

Pointer 4: The Netherlands
Ecolonia Demonstration Village Project, Alphen a/d Rijn   In August 1989 the Dutch Government approved a grant for the

building of an environmentally friendly housing project, based on the research and development work that had taken place since the Healthy Building Conference in Stockholm, Sweden, in 1988.

The results were published of 17 different architectural offices from locations throughout the country. The brief was for a 3-bedroom house which had the minimum of a 50-year life with a current (1990) cost of £30,000 excluding tax. Other stipulations were that a maximum of 1200 m³/year of natural gas would be allowed for heating, hot water supply and cooking.

Apart from energy saving the houses had to be designed to ecological principles: all materials to be 'human friendly', sound transmission between party walls to be from 0 to 4 dB. No gas cooker appliances to be allowed to emit any emissions into the room space. Good air quality and ventilation plus form, colour and natural materials to be considered.

On 15 January 1990 the location of Alphen aan den Rijn, close to Leiden (52 05 North 05 08 East), was chosen. Out of the original 17 architects nine were chosen to make final plans based on their ideas under the supervision of Belgian architect Lucien Kroll. Building construction was to start in November 1990.

Source: Novem (Nederlandse Maatschappij voor Energie en Milieu BV), May 1990.

Comment   This development through the financial initiative and confidence of the Dutch Government will become the first pioneering demonstration project incorporating all the various strands of both energy engineering and biological building methods currently available.

6 Healthy building codes and outside – inside: a storyboard ECHOES design method

Research institutions and professional bodies, particularly in Scandinavia, Denmark, the USA, The Netherlands, and most recently, the UK, are applying themselves to creating a new design and working philosphy to ensure that we can all enjoy both a healthy outdoor as well as a healthy indoor climate. Evidence of similar concerns and action is now making itself plain to the general practitioner from the Soviet Union.

Much of this work concerns the establishment of scientific criteria based on the chemical-mechanistic models laid down during the first industrial revolution. The quest for magic bullets is exemplified by the discovery of the antibiotic penicillin which revolutionized world medicine. Further quests for such examples of modern scientific magic have led to ways that can alter human genetic material. Is this to produce a human being better equipped to withstand the increasing toxic, radioactive and chemical pollutions that now surround us? It is to be hoped not, for such a path would divert us from remaining true to our natural selves.

Technological medicines to help us cure many of the cocktail of contaminants referred to in this book are of course welcome; but such a welcome must always be undertaken with care. We have looked at the problems of pollution in air, water and soils. Reference has been made to the persistence of chemicals in the food chain. Slowly we are all becoming aware that another problem of great magnitude is the vast changes taking place in the insect and animal pest populations due to the built-up resistance in such pests to the multiples of different pesticides used without care or understanding.

To date there are over six hundred species of insects known to be resistant to the pesticides used to control them; and the number is growing. It has been found that there is no one magic chemical, or one magic bullet or even one magic predator, or even one magic technique. We have to think about using many techniques in some form of synergy. Above all, we must find ways and means to achieve a natural balance of all the many forces that surround us.

In direct contrast, but for much of the time in sympathetic harmony, are the newer 'alternative' ways of thinking and doing things.

Developments in natural healing processes are under way, under the direction of Dr Vitali Pavlok, at the Soviet Scientific Commercial Medical Centre, Simferopol, Crimea, where aromatherapy is practised. The medical use of aromatic oils has been long understood by the Chinese, in particular with the reducing of stress and also respiratory illnesses. The use of in-built aromatic fragrances in building materials and furnishings is now being developed and marketed as a way of combating the illnesses resulting from chemical outgassing.

Naturally enough, new words and phrases are being coined to describe the new synergy of interaction. Biological architecture has been nurtured in Germany, Norway and The Netherlands. Ecological, organic concepts of the way in which we build our cities and buildings, and the way in which we engineer them have an essential place and may even have a greater importance than the older established set of values. But we should not allow ourselves to be mystified by new priesthoods that seem to mix yoga and architecture, eastern symbolism and a whole medley of spurious claims. Each and every challenge to the existing order of things must be thought through. Accepted that there could be some creative ingredient that we should not throw away.

One such element that has hardly been dealt with by architects and engineers is the subject of electromagnetic radiation and force fields. The Director of the technological company Rosseco (Ecology of Russia), Alexander Ivanovich Mikulin, has made the following contribution to the book:

'Apart from the external and internal factors of influence on people there is also the major factor of a distribution and level of energy fields forming from both building materials and the surrounding environment. According to observations of Soviet ecologists and parapsychologists every material of nature forms around itself its own energetic field which is different from the immediate surround, and disagrees with other magnetic and electric fields of force. The organism of man is itself not only a biological system, but is also closely associated with energy information systems. Such systems are having a profound, and as yet not fully understood, effect on the internal organs of man. Such external energy information flows come from materials of buildings (e.g. concrete), the geological zone of the country, and electromagnetic wave sources. The interaction of these uncontrolled forces upon the natural electrical forces within man himself can result in disorders to man's own internal organs with unfortunate results. Man requires protection within his buildings, and to achieve ecologically healthy buildings the following must be investigated:

1 The analysis of energy influences and spectrums from the building materials considered for use must be made to ensure that there is compatibility with man, i.e. the use of natural materials.
2 The analysis of all energy fields and electrical/power source routeways in the building construction and also external to the building.
3 To establish the position of a building so that it does not interact with natural geopathic stress points.
4 To establish analytical methods for all building materials in respect not only of energy used in creation, but also to show whether the material is low in radiation and other contaminants that can affect man.
5 To develop ways and means of shielding a building from external high-frequency electromagnetic wave influence and other energy information networks so as to create safe levels.
5 December 1990'

Alexander Mikulin gives emphasis to the information already set out on EMF factors of influence in this book. His statement shows that we are at the tip of another ecological iceberg. A time-bomb may be a more apt description, but yet another facet that we cannot ignore.

An interaction is taking place. A combination of guidelines and lines of sight is given as a basis for some tentative healthy building codes.

The use of such codes must, of necessity, relate to the actual factors of influence found on a particular project. This section ends with some examples of the use of the ECHOES (environmentally controlled human operational external/enclosed space) portrayed in a storyboard form.

Lines of sight for healthy living buildings___

The following list is a composition of methods used and tested by architect Renz Pijnenborgh, 's-Hertogenbosch, The Netherlands; architect Christopher Day, Brynberian, Crymych, Wales; and engineer Bill Holdsworth, UK and The Netherlands.

1 The building spot must be geologically undisturbed.
2 The house should be well away from motorways and industrial areas.
3 Make use of decentralization and free-living building methods in 'green districts'.
4 House and district to be individual, human and friendly; community to be stimulating, whether existing or new community.
5 Natural building materials should be used.
6 Promote natural regulation of air and humidity by the use of hygroscopic materials.
7 Encourage filtering and neutralizing of harmful materials in the air by use of absorbing materials.
8 Balance should be maintained between internal warmth and insulation.
9 Design for optimization of room space and air temperature and humidity.
10 Design for optimization of internal comfort by use of solar passive devices for heating or cooling.
11 Make minimal use of synthetic prefabricated building components.
12 Remember the maxim that a reduction of toxic emissions equals a good 'smell'.
13 Employ natural light, good illumination by natural methods and good colour sense.
14 Ensure isolation from noise and vibration.
15 Ensure elimination of any radioactive materials in the materials used for building construction.
16 Design for isolation from electrical force fields and to reduce psychological effects in living spaces (ionization).
17 Ascertain that there is no disturbance of the natural magnetic fields.
18 Ascertain that there is no disturbance of technically induced electromagnetic fields.
19 Ascertain that there is no disturbance of cosmic and terrestrial influences on the human life form (DNA).
20 Make use of natural knowledge and common-sense applications (for furnishings etc.).
21 Take care of harmonic measures, proportions and forms.
22 Use ecological materials, i.e. those that have low energy use in production and are non-polluting.

23 Ascertain that there is no removal of important raw materials from the ground, and no damage to the natural elements caused by foundations. Look at previous ground use.

International designer guidelines

The following designer guidelines are a result of research from actions taken in eight countries, including a few tips from personal experience.

1 No smoking should be allowed during construction and occupancy (for specialist clients arrange for naturally ventilated smoking rooms). NB: now a legal requirement in The Netherlands, 1990.
2 Do not use unfaced fibreglass insulation and other similar materials.
3 Carefully specify fast-drying paints, glues and mortars.
4 Use wherever possible solid hardwood (oak, birch) from registered sustainable forests for all flooring, shelving and cabinet-making.
5 Make minimal use of wallpapers, unless checked for outgassing agents.
6 Use natural materials for upholstering and furnishings.
7 Increase fresh air for ventilation, utilize heat pumps and solar devices for energy saving.
8 Use daylighting methods. Restrict the use of overhead electric lights. Use stimulated fluorescent lamps, uplighters, and wherever possible controllable task lights, as well as task-positioned personal ventilators.
9 Increase air supply to copy rooms. Check that computer equipment is not carbon monoxide emissive. Also have all copy printers in a separately ventilated room (now law in Denmark, 1990).
10 Install thermostatically controlled attic/roof exhaust fans.
11 Ensure night flushing of premises of air.
12 Increase maintenance procedures for healthy building operation and ensure that both architects and interior designers do not limit space for building services (the 'quart in a pint pot' syndrome).

Pointers towards healthy buildings of the future

From the Healthy Building report of the Nordic Seminar, March 1987.

Building location and local climate
1 Select a site which has favourable basic conditions (look out for soggy ground, risk of radon and land subsidence).
2 Place the external air intake of the building so that the quality of the air is not affected by such factors as roads, parking lots, industries, etc.
3 Orientate the building in relation to sunshine, wind, the external environment and the need for contact with the immediate surroundings.

Constructional physics and constructional engineering
1 Take steps to keep the building dry.

2 Water must be led off wherever it may occur: foundations, bathrooms, window openings, outer walls.
3 Ventilate those parts of the structure that are exposed to damp.
4 Avoid risk solutions: horizontal roofs, slabs on ground with overlying insulation, joisted floor on slabs on ground, floating floor on slab on ground.

Climate engineering

Ventilation must have a certain excess capacity to allow for human errors. Make sure that:

1 Pollutants are taken care of at source (by encapsulation, spot extraction, etc.)
2 The technology is simple and flexible to allow for changes of use of premises; to be individually controllable and to be comprehensible to the user
3 Windows are openable
4 The system is simple to inspect (fixed measurement points with finely adjusted clean and replacement components)
5 The systems are decentralized and symmetrically constructed with a high air exchange and ventilation efficiency; also without low-frequency noise
6 Ventilation systems are balanced with heat-recovery systems that do not pollute.

Avoid risky solutions such as:

1 Recirculated systems
2 Natural ventilation systems (insufficient capacity, no channelling to individual rooms, draughts)
3 Exhaust air systems
4 Air humidification
5 Hot-air systems (especially with no or minimal fresh air) spreading pollution
6 Rotary heat exchangers that spread pollution
7 Heat exchangers that cannot be turned off in summer
8 Insensitive or hypersensitive control and regulation components
9 Ventilation ducts in flooring structures

System make-up and design aspects

1 Make all systems easy to clean, maintain and run.
2 Make all systems simple, controllable, comprehensible, permanent, flexible and 'forgiving'.
3 Make sure that the end user is able individually to regulate the climate and external air flow.

Building materials

1 Use known and low-emitting materials.
2 Ask all manufacturers for a statement on pollutant emissions.
3 Make sure that materials are stable, permanent and durable for the prevailing conditions.
4 Make sure that materials do not contain heavy metals, asbestos or biocides.
5 Avoid large-surface materials such as wall-to-wall carpeting in public premises (high fluff-factor).
6 Avoid materials which may be suspected of containing toxic substances in adverse concentrations.
7 Avoid plastic wallpaper and painted glass-fibre fabric in wet rooms.

8 Avoid flooring materials which entail a personal static charge of more than 1000 V at 22°C and 25% r.h.

9 Avoid agents which protect against biological degradation — design the building so that these agents are unnecessary.

Maintenance and administration building process

1 Take into account maintenance and administration aspects of the building process.

2 Plan for careful execution, including time for drying out, fine adjustment, functional inspection and trouble-shooting (quality assurance).

3 Provide for functional and responsible co-operation throughout the entire building process from planning to occupation and follow-up.

4 Do not restrict the building to depend on sensitive technology with all the fault risks involved.

5 Include maintenance routines in the project planning.

6 Contract for long guarantee time as protection against concealed faults.

7 Complete on a qualitative rather than a monetary basis in the building process.

8 Ensure a high standard of cleaning without harmful cleaning agents.

9 Give priority to climate and hygiene aspects over energy aspects: a key lesson for operational and maintenance personnel.

10 Carry out regular functional inspections, involving the occupants.

BREEAM (Building Research Establishment Environmental Assessment Method) 1/90: an assessment for new office designs

At the time that this book was being completed some collaborative research by the UK Building Research Establishment and the Environmental Conscious Design Group of London (ECD Partnership) and others was published as the first in a series of simple design goals for the achievement of eventual healthy buildings.

The main headings are:

1 External effects
   (a) Reduction in global effects achieved by reducing greenhouse gases by high insulation, and less use of fuels that emit high levels of carbon dioxide in their production and use.
   (b) Ozone depletion by the reduction of refrigeration in air-conditioning plants
   (c) The use of wood from sustainable sources in building construction
   (d) Recycling of materials

2 Neighbourhood effects
   (a) To reduce Legionnaires' Disease, the omission of wet cooling towers
   (b) Consider local wind effects
   (c) Reuse of an existing site

Relationship between fuel use and CO_2 in UK

Fuel	CO_2 kg/kWh delivered
Electricity	0.832
Coal	0.331
Petroleum	0.302
Gas	0.198

Fig. 6.1 Aspiration cooling tower for the Drummond Gate Complex, Pimlico, London (1978). 'No noise, no water spray, safe — the cooling tower became a bold item of industrial art — set in an open piazza.' Design engineer, Bill Holdsworth; Sculptor, Sir Eduardo Paolozzi; Architects, Whitfield Partnership. Photograph by Colin Westwood.

3 Indoor effects
   (a) Protection from Legionnaires' Disease in domestic hot and cold water systems by designing to the new CIBSE code TM13
   (b) Lighting where fluorescents have high-frequency ballasts
   (c) The complete elimination of or reduction in the use of hazardous material
   (d) Improvements in indoor air quality

The first BREEAM is to be welcomed, but its limited range of assessments is not only due to the author's statement that insufficient evidence is available, but also BREEAM 1/90 reflects a political holding operation on the part of the Government then in power. The privatization of the electricity supply industry is based on the use of more, not less, electricity. Current methods of production show that it produces a higher carbon dioxide emission than coal, petroleum or gas. There is a lack of proposals for the use of alternative renewable energy resources. Having listed the elimination of hazardous materials as a primary aim, the document gives no list of safe materials to use.

 Considering that BREEAM 1/90 is for new office design there are major omissions in respect of the use and positioning of modern office-copying and word-processing equipment with their attendant hazards of which much is already known (see Section

4.2) and the problems associated with Repetitive Strain Injury, as described below.

The writer of this book recognizes such difficulties, but wonders why there is such a lack of information, when the Building Research Establishment and all the other contributors must have a greater network of information available. It is to be hoped that the compilation of future assessments will not become an issue of playing at green politics.

With a degree of modesty, may I propose to users of the BREEAM 1/90 method that many of the Healthy Building Codes and design pointers included in this book are added to expand the design assessment.

RSI (repetitive strain injury)

Writing this book was a whole new learning process, not just having to cope with the whole order of compiling, editing, indexing and collating illustrative material, but having to learn and master the art of a personal computer word processor. It was bad enough getting tired from working with an old-fashioned typewriter. But trying to find the right posture to tap out on a keyboard, and glance at a visual display unit, caused much vexation. Knowledge of positioning for natural daylight (north-side); the use of a task lamp; and periods of rest helped. The use of an old Captain's chair — my favourite — with a large pillow behind my back was not the best recommendation. Unlike full-time working journalists in their workplace office, I was able to control my working shifts.

Repetitive Strain Injury is a new environmental illness. Like the strange illnesses that Dr Sherry Rogers diagnosed in her work with outgassing, early complaints of upper limb disorders among keyboard users were treated as malingering by management in many companies.

Since the mid-1980s cases of journalists' legs have been increasing. The experience is traumatic. In addition to not being able to work, many have been unable to drive, clean or maintain their homes and gardens, cook or play any sport. Some people have been unable to wash their hair, or turn on taps.

Ways to prevent RSI have been published in a code of practice for keyboard work by the Australian National Occupational Health and Safety Commission in 1989.

Ergonomists know the answers to making the human body feel comfortable when having to remain for long periods in awkward positions. However, they may also need to see that there is a cocktail effect from many other workplace influences. Good ventilation, proper daylighting, correct placing of offices

1. Avoid direct + reflected glare. Uplighting of ceiling with special task lights recommended.
2. Do *not* have screens facing the windows. Diffuse sunlight.
3. Chair back ajustable.
4. Chair height adjustable.
5. 5 leg chair with castors for stability.
6. Foot rest to prevent pressure on back of thighs.
8. Ensure that screen can be pushed away to a safe distance and still comfortable to read.
9. Screen + keyboard — separate.
10. Screen at comfortable height and at correct angle.
11. Screen characters easy to read — adjust for brightness + contrast.
12. Allow sufficient leg room.
13. 720 mm ideal desk height.
14. 700 mm to top of desk.

Fig. 6.2   A well-designed workstation. (Source: National Union of Journalists.)

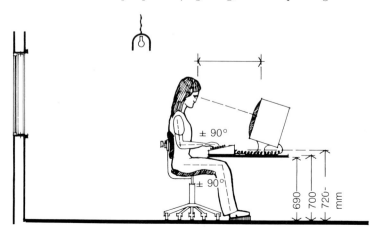

within a building, the importance of colour, introduction of plants and the absence of polluting materials can all be beneficial towards eliminating the illness of RSI.

A well-designed work station is available, based on the work of English journalist Alison Rowe, a qualified ergonomist and the Midlands UK organizer for the National Union of Journalists. Unfortunately, such well-designed work stations and well-designed ofices are still a small percentage of total office building in the majority of countries. With journalists in mind an interesting example can be found in the British House of Commons.

The office of the Press Association measures about 4 m across by 10 m long. The room is reasonably high. Narrow gothic windows allowing in very little daylight overlook an internal yard. The fluorescent overhead lights never seem to be switched off. Six to seven journalists sit among banks of computers; the screens are always close up to the operators' faces. Other electrical machinery throws off high heat energy. The room temperature is always high. There is no air conditioning. Chairs and tables are old, and totally useless for the job in hand.

In the Mother of Parliaments, where it is hoped that legislation will be passed to introduce healthy codes of practice, members of the Press responsible for communicating such information are themselves victims of bad practice.

There are six prime factors that should be taken into account when designing modern office spaces. These are:

1 Seeing
2 Hearing
3 Smelling
4 Equilibrium
5 Skin senses
6 Kinaesthetic functions (the way we sit etc.)

Healthy building (HB) codes

HB 1   The building must take nature into account. Where possible the existing vegetation should remain, and new trees, shrubs, grassland and water should be replaced in a natural, harmonic sense.

HB 2   Planning legislation must include provision for constant environmental monitoring of all discharges, and for such discharges to be reduced at source.

HB 3   National Governments should as part of planning codes of practice and building regulations, monitor, record, and regularly update pollutants affecting the outdoor environment on a regional, sub-regional and district basis.

HB 4   Recommendations for buildings close to EMF sources should include:
• Sufficient distance to be allowed between the building and the EMF source
• Provision for trees, thick bushes and earth beaming
• Sleeping areas to be over 1 m from source

HB 5   Establish a radon diagnostic list similar to that of the US Environmental Protection Agency, 1987 (see diagnostic list at end of codes).

HB 6   Establish a cancer register to make fact-finding studies more viable.

HB 7   Noise reduction measures:
- earth beams
- noise barriers
- design shape and form
- protective belts of thick trees (coniferous) and thick bushes
- heavy and porous building materials such as loam
- turf roofs
- sensitive planning to source of noise

HB 8   For checking moisture content in timber a timber meter is as important as a fever meter is to a doctor.

HB 9   Dirty water is one of the main factors in outbreaks of Legionnaires' Disease. Designers, contractors, water treatment engineers, plant managers and maintenance personnel must upgrade their attitudes to care and conservation.

HB 10   It is the job of every team member of the designing group to check every aspect of the building environment in respect of materials of construction, the system of comfort and the way in which spaces are furnished to ensure minimum indoor pollution.

HB 11   Collaboration between the architect, the structural and HVAC engineers, the client, occupiers and specialists in environmental medicine and health is essential.

HB 12   Volatile organic compounds (VOCs):
- Use maximum outside air ventilation during and following the installation of finishes and furnishings to reduce levels of any VOCs emitted from products and materials that are not free from such compounds.
- Use temporary exhausts (through doors, windows that can be opened, stair towers and emergency exits) for exhausting air, rather than the HVAC system, where it can remain. It is important to operate ventilation systems for 24 hours a day, 7 days a week, during periods of elevated VOC.
- Protect installed materials with vapour barriers where feasible during the use of finishing products with high VOC concentrations such as adhesives and paints, and during the installation of VOC-emitting furnishings and partitions.
- Operate newly occupied building areas at the lowest temperature acceptable to the occupants. High temperatures can cause bursts of VOC concentrates that have low boiling points.
- Avoid the use of fibre-glass lined HVAC ducts. Avoid the use of spray-on insulation (unless proved to be low on outgassing) for return air spaces and acoustic chambers.
- If possible do not use materials with high VOC factors.

HB 13   Electromagnetic radiation prevention code:
- Check if radon emission from ground or building materials is present.
- Check electromagnetic fields from overhead or nearby power lines.
- Check for underground water courses.
- Wrap all electrical power points and light points with aluminium foil to reduce the level of EMF.

Radon diagnostic list

The following diagnostic list was compiled by the US Environmental Protection Agency in 1987. Decision-makers, architects, engineers, builders and other designers may well find that the standards of regulatory testing in their own country fail to meet the stringent standards set out — in which case, it is proposed that they become a major item in any healthy building legislation (see HB 6).

Phase 1: initial problem assessment

The following diagnostic techniques can be used for existing sites and structures, or for new construction sites.

1 Mobile radiological van for radiation scans outdoors and indoors. These gamma scans can also employ hand-held sensors in a walk through to seek detailed local radon information.
2 Mapping of soil gas concentrations.
3 Mapping of soil permeability.
4 In structures: visual inspection with questionnaire on building and house occupancy/determine radon concentrations in the different building zones.

Phase 2: pre-mitigation diagnostics

1 Take trace gas measurements to evaluate air exchange rates and flows between zones.
2 Take a natural condition grab sample in each zone and in holes drilled into subslab, hollow-block walls, or through any other membrane suspected as a radon source.
3 Repeat under mechanical method of 10 Pascals depressurization (blower door test).
4 Carry out a blower door test of the whole house/building to evaluate building envelope tightness. (NB: the blower door test is relatively quick and inexpensive. A large fan or blower is mounted in a door or window and induces a large and roughly uniform pressure difference across the building shell. The airflow required to maintain this pressure difference is then measured. The leakier the building, the more the airflow is necesary to induce a specific indoor–outdoor pressure difference. This method can be used independently of weather conditions. ASHRAE Fundamentals 23 November 1989.)
5 IR scans of each room.
6 Check pressure differentials in basement and points of pipe/duct/cable entries.
7 Check pressure differentials with internal driers, furnace fans, ventilation equipment, etc. on/off.

Phase 3: mitigation installation diagnostics

1 Carry out trace gas evaluation of leakage to exhaust.
2 Adjust dampers for balanced airflows.
3 Check the balance by measuring pressure differentials and the velocities at key points.
4 Make an integrity check of any building envelope components penetrated by mitigation systems.

Phase 4: mitigation diagnostics

1 Determine radon concentration in various building zones.
2 Determine fan speed required for efficient radon removal; consider energy aspects.
3 Check furnace draught for airflow direction to ensure that combustion products are exiting properly.

Outside–inside: a storyboard ECHOES design method

Bill, I want nature to enter into this building. I want nature to touch all the walls, floors and ceilings. I want people to be able to feel it in all the materials and surface textures; in the quality of the air, light, and sense of well-being. Then I want nature to flow out again, undisturbed by all your pipes, electrical and mechanical things . . .

Architect Sir Denys Lasdun during a briefing for the Charles Wilson Building, Leicester University, 1966

Two examples of the ECHOES design method are presented in Chapter 4 (Section 4.1) and Chapter 5 (Case study 1).

The Staff Housing Estate for British Airways personnel at Stanwell, Middlesex in 1972 primarily concentrated on external noise from low-flying aircraft taking off from the nearby tarmac of London's Heathrow Airport. Other factors of influence were assessed and included in the design proposals but, unfortunately, were not implemented. In 1972 there was little general information about ground conditions harbouring toxic properties or radon. Neither had the general practitioner heard of electromagnetic force fields or outgassing from materials of construction.

The same was true in 1980 with the scheme for the HLH International Headquarters Building in Johannesburg, South Africa. However, many more factors of influence were identified and included. The combination of solar passive and active energy collectors for heating and cooling, the use of flowing water around the external balconies, use of an atrium and the recycling of waste water, were all methods presented by the architect Tielman Nicolopoulios and the writer, from the step-by-step methodology of the ECHOES method of walking through the building concept and using the synergy. That method was actually realized some 10 years later by architects Alberts and Huuts for the Amsterdam Bank, as described in Case study 7, p. 95.

The addition of solar passive glassed-in balconies to residential inner-city buildings, as shown in Case study 6, p. 94, coupled with the thermolabyrinth concept of filtering external air through concrete and stone earth mazes, as described earlier from examples in Germany and Poland, can give to many old buildings free-energy warmth from both indirect earth heat and direct sun heat, as well as both improving air quality and giving a greater spatial sense of living.

It should not be beyond the wit, imagination, skill and determination of architects, engineers and other prime movers to couple many of the techniques described in this book to

'Badly designed, badly built, badly managed, unhealthy desperate places in which to live. Le Corbusier's dream of "modular man" became a nightmare.'

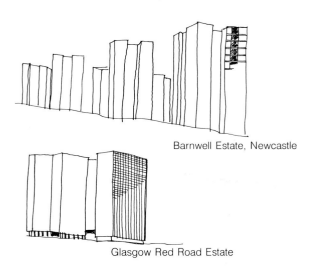

Barnwell Estate, Newcastle

Glasgow Red Road Estate

Fig. 6.3 Toxic high rise (photograph by Simon Holdsworth).

produce some drastic modifications to the inhuman scale and environmentally unhealthy concrete tower blocks that disgrace cities as far afield as Paisley, Scotland, and Moscow and other cities of the Soviet Union.

Can such concrete canisters be recycled? Is the human endeavour and cost too much? If such buildings are not wanted, then what becomes of the debris with its fingers of steel? The reader should look upon such a project as an example to solve.

As an illustration of how to use this healthy building and ECHOES method, a proposal in storyboard form is given for a Super Sports Stadium to be built in time for the 1998 World Football Cup Competition.

Preamble

In 1980 the author wrote to every First Division Football Club in Great Britain, reminding them that all fossil fuels were both finite and subject to the whims and impact of international politics. The letter pointed out that for 20 years the University of Sheffield had recorded that London obtained an average of 1000 kWh/m^2 of horizontal surface per year; that the Shetlands has the same amount of solar gain as southern France, and Dublin twice that of Milan; and an energy/cost saving was proposed for the heating of football stadiums by the use of sun energy.

Including some Second Division Clubs, over 40 were written to. Aston Villa, Nottingham Forest and Manchester United put the proposal on file. Cardiff City, Arsenal, Luton Town and Watford showed immediate interest, only later to fade.

The proposals also showed how stored solar energy could be used for year-long activities, benefiting both the community and the club.

1985 saw large-scale soccer violence. The newspaper proprietor Robert Maxwell announced his intention to expand and modernize the premises and stands at the Oxford Football Club. The 1980 proposals were re-examined and the author proposed that the ECHOES method could introduce healthy energy technology to help in the positive regeneration of community activity based on football clubs which in themselves command such great loyalties; and he further suggested that from such an amalgamation much-needed job opportunities would arise. However, the proposals for a combination of solar/phase storage and communal well-being were not taken up.

In June 1989, The Prince of Wales, who understood the link between the Hillsborough disaster and ecological balance, expressed his best wishes for such a concept.

Football stadiums (UK traditional)

1 Large open arena with open or covered stands on four sides
2 Once-a-week capacity crowds; the ground used for practice
3 A club room, small administrative offices, shower and changing rooms, 'bog' standard toilets

It should be noted that such football stadiums as those described above are the norm for many countries, not only for the UK. More dynamic behaviour could be found with the tent-like tensile structures conceived by Frei Otto in the early 1970s, with the elasticized coverings of the Hoechst Stadium, Hanover and the Berlin Stadium. It was during this time that Frei Otto was in association with Kenzo Tange in the creation of a sports

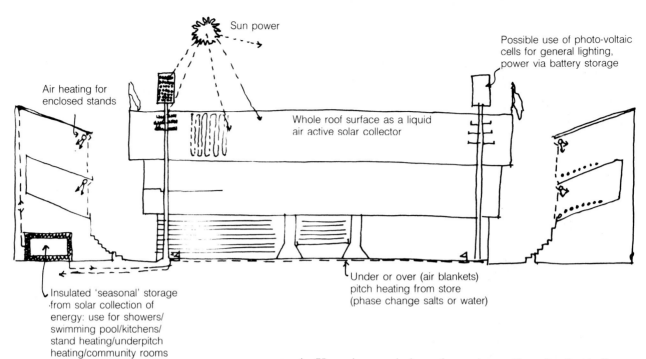

Fig. 6.4 Proposal for football club's solar heating/cooling source (sketch by Bill Holdsworth for 1980 report).

centre in Kuwait, consisting of a main stadium for football, an indoor arena and swimming pool. An organic solution was achieved to enable people to enjoy the sports in the severe climatic conditions of the location. Architectural and engineering harmony was sought to provide oases of shelter.

Football stadiums (for an 'awakening Britain')

1 A football arena with multiple facilities and use
2 An international swimming pool
3 Relaxation centre, sauna, dip pools, health baths for all ages
4 Dance hall/film—cinema centre/bowling and games centre
5 Elderly and disabled centre with provision for such people in all other areas of the complex
6 Theatre space, a community club and hobbies studios
7 Surrounding gardens, parkways, places for outdoor games, walks and simply for people to sit down in a clean environment
8 Silent, electric city trams and travelators from remote or underground car parks

Such multiple structures need not become the giant concrete expressions of brutalism that seem to echo the last 30 years of sports architecture. Some exciting possibilities were described in the August issue of *Architectural Design* in 1977 by Susan P. Gill who had been working with Frei Otto at the Institut für Leichte Flaschentragwerke at Stuttgart.

A combination of solar energy collection and environmental control was being developed with lightweight structures, materials of construction which in themselves are less energy intensive, and by their application create energy conserving opportunities.

Solar tents and solar pneumatic structures

Tent structures are particularly suited for solar energy applications because of their flexibility, adaptability (material change), convertibility (form change) and mobility. Both active and passive solar collection is possible for tent-like and pneumatic structures.

Passive solar collection and general environmental control involve recognition of the distinction between the building envelope as a barrier filter and as an energy filter.

Fig. 6.5 Earth beam protection.

We have become aware that as the technology for mechanical control of interior climate has developed, energy consumption has increased and an increasing separation between the internal and external environment has become apparent. Technology's capability for interior environmental control led to the concept of the building envelope as a barrier between indoors and outdoors. An opposite view is to consider the building envelope as a filter.

Where once a building with through ventilation creating a natural flow of air was necesary, contemporary technology has permitted such design considerations to be supplanted by mechanical systems. The importance of the position of the building relative to the sun and the context of the site was often forgotten. Climatic factors, such as local air temperatures, local wind conditions, orientation, shade and, where possible according to scale, earth beam protection, were also neglected.

Susan Gill's energy-filter concepts in 1977 as related to solar tent structures are illustrations of dealing with the total building as an ecological dynamic element, leading both to energy saving and the creation of a healthy building. Their application for many different buildings in diverse climates is radical, fresh, and apposite to the reasoning of this book.[1]

Solar tent structures can be ideal for sports and football stadiums and a number of principles are illustrated with my debt to Susan Gill's simple and effective drawings.

Three basic heat transfer principles must first be recognized:

1   Radiation
Solar collectors receive heat energy from the sun, taking advantage of the greenhouse effect by passive capture and storage

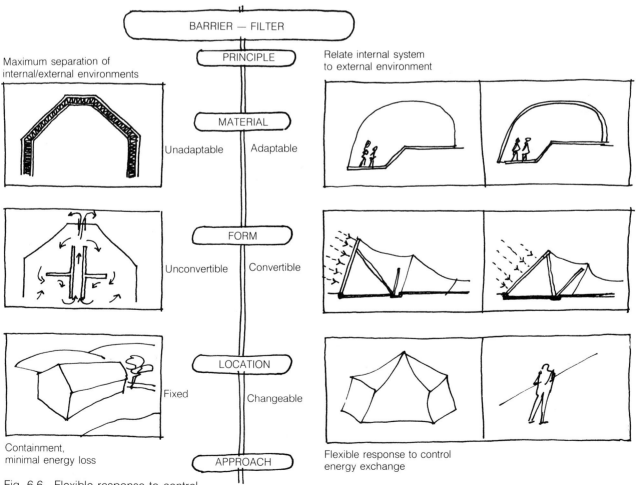

Fig. 6.6 Flexible response to control energy exchange (after Susan P. Gill).

(a) Greenhouse — transitional zone

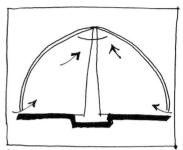

(b) A tent whose geometry promotes natural convection

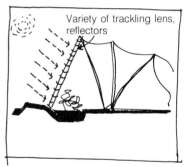

(c) Mast supported solar wall

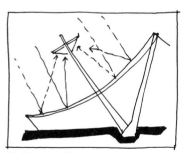

(d) Entire roof as a parabolic reflector for solar storage

(e) Lens collectors (solar cells) on masts, swivel to track the sun's path

Fig. 6.7   Solar tent structures.

of the sun's energy in the transitional zone between the indoor and outdoor areas; they also serve both in winter and summer as climatic insulators.

2   Convection
With air- or water-inflated pneumatic structures heat energy can be collected in both flexible and fixed structural envelopes.

3   Conduction
For heat transfer through a solid conducting medium the concept of the building envelope as a selective filter is important. On a winter's day when the local air temperatures are low but the sun is shining brightly, the building can be so designed that the solar heat is absorbed (by radiation) while the heat loss through the building skin (by conduction) is minimized. In such design considerations lightweight building materials with optical (for solar radiation transmissivity) and insulating characteristics are important.

Such energy collection could be by the use of a solar-heat wall integrated into the envelope of a large tent structure. The insulated heat wall could support a variety of tracking lens/reflector combinations, and the high-grade heat energy obtained could then be stored for later use for heating the open ground pitch and spectator stands, and along with heat pump principles, earth storage, etc. could provide the energy for heating, cooling and domestic hot water for all the other functions of a multi-purpose building use.

Tent-like structures can also have integrated within their skin photo-voltaic cells for the collection of sun power converted to electrical current for lighting.

Essentially, the fundamental theme of using the ECHOES design method is to explore and integrate many different concepts of energy collection, low energy and non-polluting materials.

An ECHOES football stadium

Reversing the trend, cars would be parked on the external perimeter. Covered parking could afford planted earth banks as well as places for solar collection for the provision of low-cost energy to nearby houses and other buildings (a quid pro quo for the effect on the environmental surroundings). Car exhausts could also be collected and converted to provide fuel for adapted gas-turbine engines.

The next zone would be parkland, both as an immediate amenity for the neighbourhood, and also for sports activities. Visitors to the stadium and its associated buildings would either walk, or travel on solar-operated or electric-battery-driven vehicles on tracked ways.

The inner zone would be the stadium and other buildings using all the ecological elements of design as given under factors of influence.

The scale can of course be reduced, so that the local football club can also provide similar community-based functions. Often such local football grounds are situated in built-up areas. It is in such contexts that tent structures can be viable, since the covered space, which can be made openable, can also contain areas of vegetation and green walkways and play spaces.

Using the ECHOES method the first step is to walk through the conceptual arrangement of the building complex envisaged by the client body. This same process is then undertaken at each phase of the design, working through the details of construction until the final form is then assessed.

Fig. 6.8   ECHOES's football stadium —
ecological zoning.

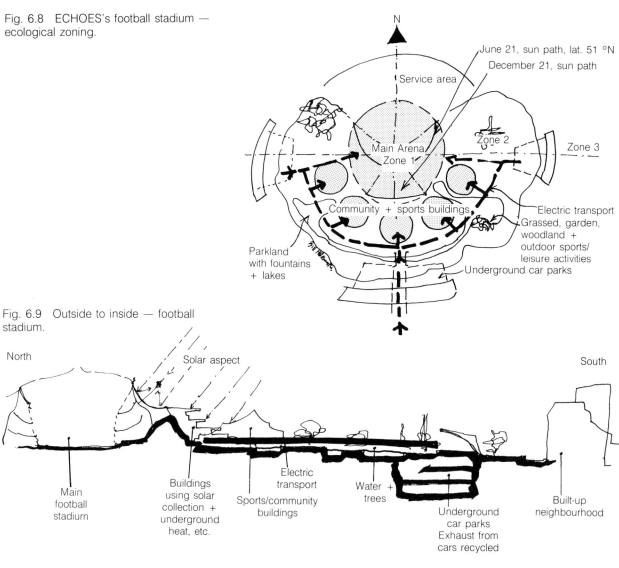

Fig. 6.9   Outside to inside — football
stadium.

	BA, London 1972	HLH Jo'burg 1980	ECHOES Stadium 1998
Factors of influence			
External			
Climate: meso and micro	*	*	*
Air quality (outdoors)		*	*
Ground conditions			*
Radiation in soil/EMF			*
Noise and vibration	*	*	*
Relationship to other buildings and infrastructure		*	*
Vegetation/visual appetizers		*	*
Internal			
Construction physics/thermal comfort	*	*	*
Daylighting/illumination		*	*
Air quality (indoors)		*	*
Radiation EMF			*
Noise	*	*	*
Building materials (volatile emissions)			*
People's needs		*	*
Technical installation	*	*	*
Energy saving (use of renewables)		*	*
CO_2 reduction and greenhouse effect reduction		*	*
Occupational behaviour		*	*
Vegetation and visual appetizers		*	*

This table shows a comparison of results from the British Airways Staff Housing Scheme, London, 1972, the HLH Headquarters Building, Johannesburg, 1980 and a conceptual Football Stadium/Community-based Sports and Leisure Complex, 1998.

7 Materials

Fig. 7.1 Are your construction methods damaging your health? (courtesy *NOVEM*, the Netherlands)

Introduction

For a building to be healthy it is essential to choose building materials with minimum pollutant emission, or outgassing, to the indoor air. A draft proposal by the Swedish Council for Building Research and the Swedish National Institute of Environmental Medicine is for materials that emit a maximum of 0.15 mg/m^3 formaldehyde and 2.00 mg/m^3 of total volatile organic compounds (TVOC).

Austria, West Germany, Denmark and Sweden have well-established manufacturers of what are being described as biological (green) materials.

The designer requires fingertip information. In many countries such immediate information is not yet available. The intention of this chapter is to help to guide the designer/specifier towards the establishment of a healthy building materials network.

Using emission data

Building designers, contractors and managers can use such data in the selection of materials, and in the operation of building services systems, especially when buildings are new, or newly refurbished, and apt to have large amounts of new surface materials with volatile organic residuals.

Manufacturers of materials can use emission data to design inherently low-emitting materials, ensure good quality control during production and develop guidelines for end-users on ways of removing residual volatiles.

One must constantly stress that many environmental illnesses (the EI syndrome) are the result of a mixture of different unhealthy emissions. Research in this field is very much at a beginner's stage.

Some chemical names decoded

Acetaldehyde An intermediary metabolite in the breakdown of some chemicals and brain hormones, also made by *candida*; highly toxic to the body if it stays in this stage and does not get metabolized further into excretable and less toxic acid.

Aldehyde A general category of intermediate metabolites that are toxic to the body if some blockage occurs.

Aromatic Merely describes the structure of hydrocarbon

Table 7.1 Organic compounds

Material/product	Major organic compounds identified
Latex caulk	Methyl ethyl ketone, butyl propionate, 2-butoxyethanol, butanol, benzene, toluene
Floor adhesive (water based)	Nonane, decane, undecane, dimethyloctane, 2-methylnonane, dimethylbenzene
Moth crystals	Para-dichlorobenzene
Floor wax	Nonane, decane, undecane, methyloctane, trimethylcyclohexane, ethylmethylbenzene
Wood stain	Nonane, decane, undecane, methyloctane, dimethylnonane, trimethylbenzene
Latex paint	2-Propanol, butanone, ethylbenzene, propylbenzene, 1,1′-oxybisbutane, butyl propionate, toluene
Furniture polish	Trimethylpentane, dimethylhexane, trimethylhexane, trimethylheptane, ethylbenzene limonene
Polyurethane Floor finish	Nonane, decane, undecane, butanone, ethylbenzene, dimethylbenzene
Room freshener	Nonane, decane, undecane, ethylheptane, limonene, substituted aromatics

chemicals and means that the carbons in chemical structure form a ring.

Benzene A common chemical; it is known to initiate leukaemia.

Candida A fungus or mould or yeast that is normal in the intestine if it grows in normal amounts; its growth is accelerated by antibiotics and sugars.

Carbon monoxide A gas in the exhausts of industry, cars and furnaces; in the bloodstream it is a poison.

Carbon tetrachloride A common chemical solvent and also a highly potent liver toxin.

Chlorpyrifos Dursban, a commonly used toxic organophosphate (inhibits nerve transmission) pesticide used in schools, homes and offices.

Cyanide A poison found in some pesticides that can affect the white matter of the brain.

Epoxides Dangerous intermediates that cause cancer and environmental illness (EI), poison the immune system, and damage genetics.

Formaldehyde A ubiquitous chemical in exhaust fumes as well as outgassing from foam products and many materials used in building.

Formic acid A metabolite of formaldehyde.

Hexane A hydrocarbon in exhaust fumes.

Hexanol The alcohol of hexane.

Mercury A dangerously toxic metal that can permanently poison many body enzymes.

Methacrylate A chemical that outgasses from plastics and adhesives.

Methylnaphthalene A potent hydrocarbon that outgasses from many household and building products and can cause cancer.

Methylenechloride Dichloromethane; a solvent in paints, paint strippers and aerosol propellant, and can rapidly metabolize in the body to a lethal level of carbon monoxide.

Mycotoxins Toxins made by moulds; many are potent carcinogens.

PCP Pentachlorophenol, a common and toxic wood preservative.

Phenol A toxic benzene derivative, the old carbolic acid. A known carcinogen.

TCE Trichlorethylene, a ubiquitous solvent and xenobiotic (foreign chemical); can affect human response mechanisms.

Toluene Methyl benzene, a solvent that outgasses from many products such as paint. It gets bio-transformed to hippuric acid in the urine.

Vinyl chloride Outgasses from plastics, can cause toxic brain symptoms.

(Source: *Tired or Toxic — a Blueprint for Health*, Dr Sherry A. Rogers, MD, Prestige Publishing, Syracuse, NY 13220, USA, 1990)

Table 7.2 Typical emission rates

Label	Source*	Condition	Emission factor**	Assumed amount	Emission rate (mg/h)
A	Silicone caulk	<10 h	13 mg/m² h	0.2 m²	3
A'	Silicone caulk	10–100	<2	0.2	<0.4
B	Floor adhesive	<10	220	10.0	2200
B'	Floor adhesive	10–100	<5	10.0	<50
C	Floor wax	<10	80	50.0	4000
C'	Floor wax	10–100	<5	50.0	<250
D	Wood stain	<10	10	10.0	100
D'	Wood stain	10–100	<0.1	10.0	100
E	Polyurethane wood finish	<10	9	10.0	<1
E'	Polyurethane wood finish	10–100	<0.1	10.0	<1
F	Floor varnish		1.0	50.0	50
G	Particle board	2 yrs old	0.2	100.0	20
G'	Particle board (HCHO)	New	2.0	100.0	200
H	Plywood panels	New	1.0	100.0	100
I	Chipboard		0.13	100.0	10
J	Gypsum board		0.026	100.0	3
K	Wallpaper		0.1	100.0	10
L	Moth cake (para)	23°C	14,000	0.02	280
Combustion sources					
M	Unvented gas burner		85–144 mg/h	1 burner	100
N	Unvented gas space heater (radiant HCHO)		0.001 mg/kJ	20,000 kJ/h	20
O	Unvented kerosene space heater con/radiant		0.007 mg/kJ	6100 kJ/h	45
P	Unvented kerosene space heater radiant		0.064	9400	600
Q	Cigarette smoking (one smoker)		10 mg/cig.	2/h	20
Activity sources					
R	Hair spray	6 sec	3 mg per use	1 use/h	3000
R'	Hair spray	6 sec	3 mg per use	1 use/day	120
S	Disinfectant spray	6 sec	5 mg per use	1 use/h	5000
S'	Disinfectant spray	6 sec	5 mg per use	1 use/day	210

Source: Environmental Protection Agency, Washington, USA.

* Emission data shown are typical only for the brands that were tested and do not represent all products. Product-to-product variability can be very high.

** Typical values selected from values presented to ASHRAE Conference on IAQ, Atlanta, USA, 1988.

Label = source type label on graph: typical emission rates in US residences, 1988.

Para = paradichlorobenzene.

HCHO = formaldehyde.

These lines correspond roughly to emission rates that in a typical dwelling of 300 m^3 and 0.5 air changes per hour would respectively lead to total house (dwelling)-wide concentrations well above and well below 1000 μg/m^3. The 1000 μg/m^3 concentration is simply a single-source maximum contribution that might be advisable based on the work of Molhave (i) which suggests that concentrations of less than 5000 μg/m^3 of total vapour-phase organics can be irritating to some people.

This is a concept to be used as interim guidance to building designers and manufacturers until more definitive data on exposure to low-level organic compounds becomes available.

Example The particular floor adhesive and floor wax B and B′ have average emission rates well above the upper band during the first 10 hours after they are applied. The aerosol disinfectants R and S are also above the upper band. Some products fall into a 'grey' zone between 100 and 1000 mg/h, and most fall into levels where they seem to be safe.

In principle, such plotting can help to identify major sources of pollutants, and give manufacturers guidelines to the suitability of their products.

But care must be advised, since it is known that such products of outgassing last longer than expected by entering other building components, and mixing with other outgassing agents. Also, there are additions from other activities and the overall levels increase. (μg/m^3 is the amount of microgram particles per cubic metre.)

Precautions with building materials and furnishings

Avoid the use of

1 Oil-based impregantion methods and polyfilla containing quick-drying agents, and fungicides
2 Uncovered mineral fibre materials
3 Glued carpets
4 Hydrochloric acid on brick walls without heavy pre-watering
5 Materials containing formaldehyde
6 Concrete products containing plastic and high-radon (blue shale) content
7 Glues, varnishes and paints containing chemical additives
8 Plastic surfaces on walls and floors
9 Synthetic carpets with underlays of PVC; these give off small quantities of a highly poisonous substance, vinylchloride monomer

Energy content of key building materials in common building units

Changes in global climate by the pollution of the atmosphere have a direct link to the methods and ways in which we use fossil-fuel-derived energy. The energy link to building construction materials is an important element in the creation of a healthy climate in which to build.

Table 7.3 is from the work of John Connaughton of the Davis Langdon and Everest Consultancy Group, London, 1990.
Note: Romig and O'Sullivan in *Buildings and the Energy Future*

Table 7.3   Energy content of key building materials (GJ/tonne)

	1	2	3	4	5	6	Average
Cement	5.7	6.1	6.0	6.8	7.8	7.8	6.7
Aggregates	—	—	—	—	—	0.1	0.1
Common bricks	3.4	2.8	—	6.3	—	3.1	3.9
Flat glass	11.9	15.0	—	—	22.2	12.4	15.3
Steel	24.0	—	—	—	37.3	31.0	30.7
Timber	6.4	2.3	—	—	—	2.7	3.8

Source: Davis Langdon and Everest, based on various sources, 1975–1981.

Table 7.4   Material energy requirements for beams of equivalent structural performance

550 × 135 mm softwood glulam	1 energy unit
Reinforced concrete beam	5 energy units
305 × 165 mm steel I-section	6 energy units

Source: Ron Marsh, Ove Arup and Partners, UK.

Table 7.5   Material and construction energy requirements for a 2200 m² warehouse

Construction specification	Energy required (GJ)
Timber throughout	1480
Concrete block walls and timber roof	2550
Steel prefabrication	3150
Concrete tilt-up walls and timber roof	4030
Steel prefabricated with aluminium cladding	4830

Source: TRADA, Timber Research and Development Association, UK.

(1979) estimated that some 56% of developed energy use in the UK in 1978 was used in buildings. The situation has hardly changed (1990). The energy consumed in the building materials industry in providing materials input to one year's supply of new buildings is more than five times greater than the energy consumed by these buildings in the first year of use (see Table 7.4).

From a study made in the USA in 1990 the energy requirements of materials and construction for a 2200 m² warehouse were as given in Table 7.5.

Building with stabilized earth

Subsoil is one of man's oldest building materials. It is available on or near most sites where people choose to erect shelters; it requires the simplest equipment to be transformed from its raw *in-situ* state into an extremely strong, durable, heat and sound-insulating, fireproof and rotproof external wall.

It has fallen into temporary neglect in the industrialized nations because of competition from far cruder materials such as concrete.

Subsoil is a composite material which contains a combination of the following elements: clay, sands (of varying degrees of courseness) silt, water, air and organic matter. The manipulation of these elements can lead to precise control of the material's constructional properties. The balance of the elements is rarely such that it can be used for construction without some modification. However, simple techniques have been developed to process just about any subsoil (from clays, to loams, to very sandy soils) so that it can be used reliably.

The process is called stabilization, which can be both hand processed and mechanically undertaken, where compaction is obtained after the right mixture of the elements has been obtained.

The most favoured bonding addition has been cement, but lime is a material much used in the USA, and in bio-architectural building techniques in Germany and The Netherlands. Other reinforcement can be undertaken with vegetable fibres, chopped straw, reeds and grasses.

Forming is undertaken in a traditional way. When it comes to weatherproofing earth walls limewashing and roughcast allows the material to be reused and is also an effective weatherproof barrier, as proved when the toughest mix has been used for lighthouses. There are also some biologically derived waterproofings now on the market.

Subsoil has an enormous unrealized potential. (In The Netherlands loam wall construction has been found to provide protection from radon penetration as well as to improve energy insulation in a cheap and effective way.) It is a subtle material that has been awaiting subtle development. It is free, it's available just about everywhere, it is simple to make and operate, its constructional performance is high and its careful use can trouble the planet far less than most alternative materials.

(The information contained in this item on building with stabilized earth is based on an article written by Colin Moorcraft: Eco-tech in *Architectural Design*, UK, January 1973.)

A fragrance for health: aromatherapy and advanced polymer technologies

Specially prepared organic polymers are now available that can entrap fragrances, flavours and other active ingredients in a patented, odourless way that can be moulded or sprayed, or used as a coating on numerous substrates including paper, fabric or metal.

Aromatic substances from plant and animal resources had been fairly widely used in medieval times for physical and psychological treatments.

In recent years re-evaluation of old Chinese drugs and 'crude' drugs has been made, probably in reaction to modern Western medicines. Increasingly studies are being done on the actions of perfume on mental states: e.g. the forest therapy conducted in Germany, the treatment of psychosomatic illness by inhalation of incense, and the treatment of dysequilibrium (dizziness) also by the inhalation of incense, as discovered by Professor Hinoki at Kyoto University, Japan.

Kohdoh, or the art of fragrance, is an established art in Japan. Many essential oils, extracts and other aromatic substances can help and cure somnolence, migraine, depression, tobacco smoking and apathy. In one Japanese experiment it was found that when the office air was scented with lavender stress-related errors per hour of key-punch operators dropped by 21%. Jasmine, which aids relaxation, made an error rate decline of

33%. According to Junichi Yagi, Vice-President of Shimizu Inc., 'Even when the scent was below conscious levels the workers reported that they felt better than they did without it'.

Aromatics is not another new-age fad. Its use dates back some 6000 years to ancient Egypt. The core of aromatherapy is the use of essential oils to aid healing.

A US company, Creative Environments Inc., with corporate headquarters and manufacturing facilities in Pennsylvania, are now world leaders in high-technology polymer entrapment. Such aromatics can be introduced into naturally healthy materials to add another dimension to the indoor air quality: the ability to be able to help people's health from the effects of our continuing poor outdoor environment.

Further information can be obtained from:

Chella Luthy, President,
Creative Environments Inc.,33 West 54th Street,
New York, NY 10019, USA
(Tel: 212 459-9710; fax: 212 459-9713)

Michiaki Kawasaki,
Takasago Corporation,
Tokyo Research Laboratory,
36-31 Kamata, 5-chome,
Ohta-ka, Tokyo 144, Japan
(Tel: 734 1211 ex.252)

Manfacturers and suppliers' guide to healthy building materials

The source information has been collected from various countries. It has been compiled as a means for the designer to establish a materials network.

The compilation has been made from information available at the time of writing.

Austria

Building Construction Materials 1988/89
(Baukonstruktionen und Baustoffe Baubiologie in der Praxis)

Austrian Institute for Building Biology, IBO
A-1030 Wien, Landstrasse Hauptstrasse 67
Tel: 0222 713 37 93

This is an excellent publication, published in German, which gives practical building examples and some 30 pages of advertisements from suppliers of healthy building materials.

Denmark

a/s Fibo, Volddbjergvej 16, DK-8240 Risskov
Tel: 45 6 17 39 00   fax: 45 6 17 39 66

Manufacturers of expanded clay lightweight aggregate, including lightweight blocks: the material has high thermal and sound-insulating properties; also, a lightweight block suitable for foundations and external and internal walls, including Fibomatt wall panels.

Clay and loam products are natural and useful to limit both radon and EMF penetrations.

Germany
The Institute for Building Biology, Holzham 25
D-8201, Neubeuern, Germany
Tel: 8035 2039

Publishes a monthly magazine, *Wohnung + Gesundheit* (in German) with a detailed products list.

The Netherlands
Thissen Kleiwaren, Meijelseweg 37, 5986 NH Beringe, NL
Tel: 31 4760 71648   fax: 31 4760 76352

Manufactures a range of hollow building blocks, flue components and prefabricated floor components from clay and loam products.

CVK ua Postbus 23, Utrechtseweg 38, Hilversum, NL
Tel: 31 35 211553   fax: 31 35 211371

In development with architect Renz Pijnenborgh a low-energy warm-wall system set in natural limestone prefabricated lightweight blocks that produces good background heating and stable room air temperatures.

Agathos, Mercuriusweg 2, 7202 BS Zutphen, NL

Manufacturers of environmentally friendly paints, wallpapers and pastes.

The Dutch Institute for Integrated Biological Architecture, VIBA (Nederlandse Vereurging voor Integrale Biologische Architectur, Den Bosch, The Netherlands), publishes a monthly magazine 'Gezond Bouwen en Wonen' (in Dutch) with a detailed products list.

UK
Cork Slabs:
Euroroof Ltd, Denton Drive, Northwich,
Cheshire CW9 7LE, UK
Tel: 0606 48222

Gypsum plaster:
British Gypsum Ltd. Tel: 0602 214161

Insulation materials:
Rockwool Ltd., Pencoed, Bridgend,
Mid Glamorgan CF34 6NY, UK
Tel: 0656 862621

Paints, waxes, resin oils and solvents:
AURO Organic Paints, White Horse House,
Ashdon, Saffron Walden, CB10 2ET, Essex, UK
Tel: 31 79984 240

These paints are widely used in Germany and The Netherlands and can also be obtained in Eire:

AURO-Naturfarben
Dr Hermann Fischer
Direktor
3300 Braunschweig
PO 1220
Germany

AURO Ireland Ltd, Fivemilbourne,
County Leitrim, Eire.
Tel: 71 43452.

8 References and resources network

References

Chapter 2 Health as a design element

1 Rousseau, D. (1988) *Your Home, Health and Well-being*, Hartley and Marks, Vancouver.
2 Morris, David (1990) The ecological city as a self-reliant city. In *Green Cities* (ed. David Gordon), Black Rose Books, Montreal/New York, pp. 20–35.
3 AIBB (1989) *Baukonstruktionen und Baustoffe Baubiologie in der Praxis*, Austrian Institute for Building Biology, Vienna (in German).
4 Cannell, M.G.R. and Hooper, M.D. (1990) *The greenhouse effect and terrestrial ecosystems of the UK*, Institute of Terrestrial Ecology publication No. 4, HMSO, London.
5 Holdsworth, W. (1972) Micro-climates and district heating. *Conference Proceedings on Buildings and Energy, Budapest.*

Other useful reading
HMSO (1989) *Occupational Exposure Limits 1989*, COSHH Regulations, HMSO, London.

Chapter 3 Climate and human life

1 Brooks, C.E.P. (1954) *The English Climate* (revised by H.H. Lamb) English Universities Press, Sevenoaks, Kent.
2 Referred to in Landsberg, H. (1962) *Physical Climatology*, Gray, Pennsylvania.
3 Lacy, R.E. (1971) *An Index of Driving Rain*, Building Research Establishment, HMSO, London.
4 Siple, P.A. and Passel, C.F. (1945) Measurements of dry atmospheric cooling in sub-zero temperatures. *Proceedings American Philosophical Soc.*, ch. 89 p. 177.
5 Landsberg, H. (1953) *Scientific American*, August issue.
6 Olgyay, V. (1963) *Design with Climate*, Princeton University Press, Princeton, New Jersey.
7 Huntingdon, E. (1963) *The Human Habitat*, Norton, New York (first published 1927).
8 Marcus Vitruvius Pollio (1960) *The Ten Books of Architecture* (trans. Morgan), Dover.
9 Penwarden, A. and Wise, A.F.F. (1975) *Wind Environment around Buildings*, HMSO, London.
10 Holdsworth, W. (1980) The building as a solar collector. *Clima 2000 Conference, Budapest.*

11 Banham, R. (1969) *The Architecture of the Well-Tempered Environment*, Architectural Press, London.

Other useful reading

Aronin, J. (1953) *Climate and Architecture*, Reinhold, New York.

Barry, R.G. and Chorley, R.J. (1987) *Atmosphere, Weather and Climate*, 5th edn, Methuen, London and New York.

Caborn, J.M. (1965) *Shelterbelts and Windbreaks*, Faber, London.

Chandler, T.J. (1965) *The Climate of London*, Hutchinson, London.

Geiger, R. (1965) *The Climate Near the Ground*, Harvard University Press, Cambridge, Mass.

Hardy, R. *et al.* (1982) *The Weather Book*, Mermaid Books, London.

Lamb, H.H. (1982) *Climate History and the Modern World*, Methuen, London and New York (reprinted 1985).

Leland, Rogers C. and Kerstetter, R.E. (1974) *The Ecosphere*, Harper and Rowe, New York.

Givoni, B. (1968) *Man, Climate and Architecture*, Elsevier, London.

Pilkington Bros. (1969) *Windows and Environment*, McCorquodale.

Sealey, Antony, F. (1979) *Introductions to Building Climatology*, Commonwealth Association of Architects, UK.

Chapter 4 Factors of influence
4.1 External

1 Clarensjo and Kumlin (1984) *Radon in Dwellings. Remedial Actions for New and Reconstructed Dwellings*, Swedish Council for Building Research, R90, Stockholm.

2 Anon (1987) Reeds that Clean the Earth. *Environment Now*, No. 1, November issue.

3 Sopper, W. (1990) Forests as living filters for urban sewage. In *Green Cities* (ed. David Gordon), Black Rose Books, Montreal/New York, pp. 145–58.

4 DHEW (1988) *Geomagnetism, Cancer, Weather and Cosmic Radiation*, US Department of Health, Education and Welfare, Salt Lake City, Utah.

5 Konig *et al.*, (1981) *Biological Effects of Environmental Electromagnetism*, Springer Verlag, New York.

6 *Biologisch Architekten Kollektief*, Amsteldijk 21, 1071 HS, Amsterdam.

7 Joachim Berendt (1988) *The Third Ear on Listening to the World*, Element Books, Dorset.

8 Raza, Murthy *et al.* (1988) Ability of plants in development of healthy urban colonies. *Proceedings of Healthy Building Conference, Stockholm*, Vol. 2.

Other useful reading

Von Pohl (1987) *Earth Currents, Causative Factors of Cancer and Other Diseases*, Frach-Verlag, Stuttgart.

Smith, C.W. and Best, S. (1988) *The Electromagnetic Man*, Dent, London.

North, M. (1986) *The Vision of Eden: The Life and Work of Marianne North*, Webb and Bower, Exeter, Devon.

Clean Air Research Pack: Things of Science (Cambridge) Ltd, Advisory Centre for Education, 32 Trumpington St, Cambridge, UK.

4.2 Internal

1 Sanders, C.H. (1988) How to avoid surface condensation. *Healthy Buildings Conference proceedings, Stockholm*, Vol. 1.

2 Bunch-Nielsen, T. (1988) *Moisture Measurement in Wooden Building Components*, Miljoteknik a/s. Birkerod, Denmark.
3 Cobolt Arkitekter (renamed Arkitekgruppen GAIA) Jollestro, 4560, Vanse, Norway.
4 Day, Christopher (1990) *Places of the Soul — Architecture and Environmental Design as a Healing Art*, The Aquarian Press, Wellingborough.
5 Haydt, Gerhard and Dietsch, Peter (1988) *Toxic Induced Personality Changes and Deterioration of Performance after Exposure to Wood Preservatives*, Burger Hospital, Stuttgart.
6 Rogers, Sherry (1986) *The EI Syndrome*, Prestige Publishing, New York; (1990) *Tired or Toxic*, Prestige Publishing, New York.
7 Berglund, Birgitta, Report for the National Institute for Environmental Medicine, Stockholm.
8 Bunn, Roderic (ed.) (1986) *Building Services Journal*, September issue.
9 Report by the Heating and Ventilating Contractors' Association, 34 Palace Court, London W2 4JG 1986.
10 Hambleton, P., Boster, M.G., Denis, P.J. *et al.* (1983) *Journal of Hygiene*, Cambridge Press, Cambridge, UK.
11 Korsgaard, Jens (1988) The Chest Clinic, Aarhus Municipal Hospital, Aarhus, Denmark. Demands of the allergic and hypersensitive populations. *Healthy Buildings Conference Proceedings, Stockholm*, Vol. 1.
12 Fanger, P.O. (1988) The introduction of the olf and the decipol units to quantify air pollution perceived by humans indoors and outdoors. *Energy and Building*, **12**, 1—6.
13 Fanger, P.O. (1988) *A Comfort Equation for Indoor Air Quality and Ventilation*, Heat and Climate Laboratory, Danish Institute of Technology, Lyngby, Denmark.
14 Tucker, G. (1989) *Air Pollutants from Surface Materials: Factors Influencing Emissions*, US Environmental Protection Agency, (EPA), Washington DC.
15 Hult, Marie (1990) Stockholm Social Welfare Administration and Byggforskning (Building Research), Stockholm pre-publication report.
16 EPA (1986) *PCP: An Environmental and Clinical Study*, No. 560/5-87-001, US Environmental Protection Agency (EPA), Washington DC.
17 McNall, Preston (1988) Prescription for a healthy building NVCA system. *Healthy Buildings Conference Proceedings, Stockholm*, Vol. 1.
18 NBRI (1990) *Healthy Aspects of Indoor Climates (Gezondheidsaspecten van Klimaatinstallaties)*, Rijsegebouwdienst (National Building Research Institute), 's-Gravenhage (The Hague) (in Dutch).
19 ASHRAE (American Society of Heating, Refrigerating and Air Conditioning Engineers, Inc.) 1791 Tullie Circle, NE Atlanta, GA 30329 USA, March 1990.
20 Danish Technological Institute (1990) Report in *The Observer*, London, May.
21 Appleby, Paul (1989) Design guide for displacement ventilation. *Building Services Journal* UK, April.
22 World Health Organization (WHO) (1969) Magnetic fields: environmental health criteria, WHO, Geneva.
23 Robertson, Heather (1988) Criteria for selecting a healthy house in Sydney, Australia. *Healthy Buildings Conference Proceedings, Stockholm*, Vol. 3.
24 Becker, R.O. (1990) *Cross Currents*, Jeremy P. Tarcher Inc., USA.

25 De Matteo, R. (1985) Terminal Shock, NC Press, Toronto.

26 Wertheimer, N. and Leeper, E. (1986) Electric blanket/miscarriage study. *Bioelectromagnetics*, **7**, 13.

27 Milham, S. (1985) *Environmental Health Perspectives* 62 (1985:297 Epidemiological study of persons occupationally exposed to all types of man-made electromegnetic radiation).

28 Oliver, D. (1986) *Energy Efficiency and Renewables: North American Experience*, Energy Advisory Associates, Milton Keynes, UK.

29 Ashley, Stephen (1988) Plant control. *Building Services Journal*, December issue.

Other useful reading

Curwell, S.R. and March, C.G. (1986) *Hazardous Building Materials: A Guide to the Selection of Alternatives*, E. and F.N. Spon, London.

Curwell, S.R., March, C.G. and Fox (1988) *Use of CFCs in Buildings*, Fernsheer Ltd, UK.

Weir, F. (1989) *Safe as Houses? CFCs in Buildings: Insulating Foams and Air Conditioning*, Friends of the Earth, London.

Anon (1989) *The Good Wood Guide*, Friends of the Earth, London.

Bowdidge, J.R. (1989) *Pollution Reductions through Energy Conservation*, UK Mineral Wool Association, UK.

Wilson, S. and Hedge, A. (1987) *The Office Environment Survey: A Study of Building Sickness*, Building Use Studies Ltd, 14/16 Stephenson Way, London NW1 2HD.

CIBSE (1987) *Minimizing the risk of Legionnaires' Disease*, Technical Memoranda TM13, Chartered Institute for Building Services Engineers (CIBSE), Delta House, 222 Balham High Road, London SW12 9BS.

Birkin, M. and Price, B. (1989) *C for Chemicals: Chemical Hazards and How to Avoid Them*, Green Print, UK.

Potter, I.N. (1988) *The Sick Building Syndrome: Symptoms, Risk Factors and Practical Design Guidance*, Technical Note 4/88, BSIRA, HMSO, London.

Rousseau, D. *et al.* (1988) *Your Home, Health and Well-Being*, Hartley and Marks, Vancouver.

Soyka, F. and Edmonds, A. (1981) *The Ion Effect*, Bantam Books, New York.

BRE (1987) *Insecticidal Treatments Against Wood-boring Insects*, Building Research Establishment (BRE) Digest No 327, BRE, Princes Risborough, Bucks.

Weir, F. (1990) *Towards Ozone-friendly Buildings*, Friends of the Earth, London.

Kinnersly, P. (1973) *The Hazards of Work* (excellent section on toxic substances) Pluto Press, London.

Bertell, R. (1985) *No Immediate Danger: Prognosis for a Radioactive Earth*, The Women's Press, London.

Lovelock, J.E. (1989) *The Ages of Gaia*, Oxford University Press, Oxford.

Lovelock, J.E. (1988) *Gaia — A New Look at Life on Earth*, Oxford University Press, Oxford.

Day, C. (1990) *Building with Heart — A Practical Approach to Self and Community Building*, Green Books, UK.

The Conservation Society (1979) *Proceedings of Symposium on the Toxic Effects of Environmental Lead* (held at The Zoological Society, London, in May), The Conservation Society, UK.

NBHW (1987) *Tobacco Control in Sweden*, The National Board of Health and Welfare (NBHW), 10630 Stockholm.

WMO, UN and WHO (1987) *Proceedings of Symposium on Climate and Human Health — World Climate Programme: Applications,*

Leningrad (3 volumes), World Meteorological Organization (WMO), United Nations (UN) and World Health Organization (WHO).

Cottam, David (1987) *Building the Pioneer Health Centre*, Architectural Association, London.

SEE (1971) *Environmental Engineering: Aspects of Pollution Control*, papers presented to a symposium of the Society of Environmental Engineers (SEE), 68a Wigmore Street, London.

The Horticultural Research Centre, Sussex, UK is producing conversion factors for lamp lux levels to suit plants.

Engles, Joep (1989) *Radon in internal house environments (radon in het binnenhuismilieu)*, Milieu Defensie, Amsterdam (in Dutch).

EIU (1990) *Construction Materials and the Environment — Preparing for Stricter Building Product Standards*, The Economist Intelligence Unit (EIU), 40 Duke Street, London.

Becker, Robert O. (1990) *Cross Currents*, Jeremy P. Tarcher Inc., Los Angeles.

Bateman, Peter (1989) Bringing the outdoors indoors. *Turf Management*, October issue.

Hollis, Chris (1990) UPSTART: a plague of modern-day proportions. *Sunday Times*, 11 March issue.

Bateman, Peter (1989) The anti-social all-consumer. *Country Life*, 18 May issue.

Bateman, Peter (1988) Avoiding bio-slime and bafflejelly. *Food Processing Magazine*, December issue.

Chapter 5 The building as a third skin

1 CIBE Heritage Group (1989) Dr Jeffrey's earth-cooling system. *Building Services Journal*.

Chapter 6 Healthy building codes

1 Gill, Susan P. (1977) *Architectural Design*, **47**, Nos 7–8.

Endpiece Regaining a global perception

1 Coldicutt, Susan (1991) personal communication.

Resources network

British Association of Landscape Industries (BALI)
9 Henry St, Keighley, West Yorkshire, BD 21 3DR, UK.

Building Biology, Institute for
White Horse House, Ashdon, Saffron Walden, Essex CB10 2ET, UK. Tel: (079984) 240. The Institute collects information and publications (some for sale), recommends architects, specialist surveyors and test laboratories.

Building Research, The Swedish Council
Sangt Goransgatan 66, 112 33 Stockholm, Sweden. Tel: 46 8 617 73 00.

CIB W67, Energy Conservation in the Built Environment
Professor Ingemar Hoglund, The Royal Institute of Technology, S-111 44 Stockholm, Sweden. Research, publications and conferences.

Conservation Society, The
12a Guildford Street, Chertsey, Surrey, KT16 9BQ. Tel: (09328)

50975. Founded in 1966 to publicize and combat two global threats: the population explosion and the headlong industrialization of the earth.

Dulwich Health Society
130 Gypsy Hill, London SE19 1PL, UK. Tel: (081) 670 5883. Non-profit making, the society has information on geopathic stress.

Environment Conscious Builders, Association of
Keith Hall, Hopton road, Cam, Dursley, Gloucestershire, UK. Tel: (0453) 542531. Information on environmentally friendly building products.

Environmental Medicine Foundation
Dr Jean Monro, medical adviser, Breakspear Hospital, High Street, Abbots Langley, Hertfordshire, UK. Tel: (09277) 61333. The Foundation keeps records of environmental health groups for local support. The hospital is Europe's first major purpose-built environmental medicine hospital (1990).

Environmental Network, The Women's
287 City Road, London EC1V 1LA, UK. Tel: (071) 490 2511. Exhibitions, publications, information on ecological suppliers.

Ecological Design Association
David Pearson, c/o Gaia, 66 Charlotte Street, London, W1P 1LR. Tel: (071) 323 4010 ex. 204. Aims to promote ecological design, inform the public, and to promote the setting up of standards for ecological projects.

Pioneer Health Centre Ltd, The
Mrs Pam Elven, Hon. Secretary, Camolin, Birtley Rise, Birtley, Guildford, Surrey, GU5 0HZ, UK. To further the aims of Scott Williamson and the first family health centre in Britain built in Peckham, London, in 1926. Also, to promote an holistic approach to all aspects of health in the built environment.

NOVEM (Nederlandse Maatschappij voor Energie en Milieu), The Dutch National Institution for Energy and the Environment
P F M van de Laar, Communications Manager, Leidseveer 35, 3503 RE Utrecht, The Netherlands. Tel: 31 30 363401. (All enquirers answered in English.)

Ministry of Housing and Environment, The Netherlands
National Building Service (Rijksgebouwendienst afdeling Technik en Kwalitietszorg), President Kennedylaan 21, 2500 EZ, The Hague, The Netherlands. Tel: 31 70 36 14 221. Research publications are in Dutch. However, the organization will answer all enquirers in English.

Research institutes (international) and contact addresses for people working in the field of healthy buildings

Screensafe electrostatic reduction device, Per Alm, General Manager, Screensafe, Storgatan 45, S-392 Kalmar, Sweden

Australia
Coldicutt, Susan, Architect, University of Adelaide, Department of Architecture, PO Box 498, Adelaide, 5001 SA.

Austria
Panzhauser, Erich, Dipl Ing, Professor, University of Technology, Karlsplatz 13, 1040 Vienna.

Belgium
Laboratory of Building Physics, KU Leven, Kasteel van Arenburg, 3030 Heverlee.

Canada
National Research Council of Canada, Institute for Research in Construction, Saskatoon, Sask, S7N OW9.

Ontario Research Foundation, Occupational Health, 2395 Speakman Drive, Mississauga, Ontario, L5K 1B3.

Confédération des Synicats Nationales, 1601 de Lorimier, Montreal, Quebec, H2K 4M55.

China, People's Republic of
Institute of Building Physics, 19 Che Gong Zhuang Street, 100044 Beijing.

Czechoslovakia
Slovak Technical University, Civil Engineering Faculty Radlunskeho 11, 813 68 Bratislava.

Denmark
Danish National Institute for Occupational Health, Lerso Parkalle 105, 2100 Copenhagen 0.

Danish Building Research Institute, PO Box 119, 22970 Horsholm.

National Building Agency, Stormgrade 10, 1470 Copenhagen K.

Egypt
Wassef Nagui, Architect, 53, Noubar Street, Apt 22, Bab-el-Louk, Cairo.

Finland
Ministry of the Environment, PB 399, 00121 Helsinki.

University of Kuopia, Department of Environmental Hygiene, Box 6, 70211 Kuopia.

France
Barton, Alexis, Professor, Université de Rennisi, 13 Allée des Korrigans, 35510 Cesson-Sevigne.

Germany
Fraunhofer Institut für Bauphysik, Bereich Warme/Kilama, Nobelstrasse 12, 7000 Stuttgart.

Institute for Water, Soil and Air Hygiene, Corensplatz 1, 1000 Berlin 33.

Hungary
Hungarian Institute for Building Science, David Ferenz ut 6, 1113 Budapest.

Institute for Quality Control of Building, Dioszegi ut 37, 1113 Budapest.

Italy
Alfano, Gaetano, Professor, University of Naples, Piazzale Tecchio 80, 80125 Naples.

Raffellini, Giorgio, Professor, University of Florence, 1st Fisica Technia, Viale Risorgimento 2, 40136 Bologna.

Japan
Sekisu House Ltd, Tokyo Planning Office, Nishi-Shinjuku, 1-11-3, Shinjuku-ku, Tokyo 160.

Yazaki Co. Ltd, Air Conditioning, Research and Development, 1370 Koyasucho, Hamamatsu-shi, Shizuoka-ken.

Kimura, Ken-Ichi, Professor, Waseda University, Department of Architecture, 3-4-1 Okubo, Shinjuka, Tokyo 160.

Morikawa, Yasushigo, Dr, Taisei Corporation, Technical Research Centre, 344-1 Nasecho, Totsuka-ku, Yokohama-shi, Kanagawa-Ken.

Nakamura, Yasuto, Assistant Professor, Kyoto University, Yoshida-Honmachi, Sakyo-ku, Kyoto 606.

The Netherlands
Schmid, Peter, Professor, Eindhoven University of Technology, PO Box 513, 5600 MB, Eindhoven. (Professor Schmid has written many papers for international conferences, and is a leading exponent of building biology in The Netherlands.)

The International Institute for the Urban Environment, Nickersteeg 5, 2611 EK, Delft, The Netherlands. Tel: 31 15 62 32 79. Director: Tjeerd Deelstra.

Stichting Mens & Architectuur, p/a Amsteldijk 11 a, 1074 Amsterdam, The Netherlands. Tel: 31 02 67 11 734. Secretary: Marleen Kaptein.

Switzerland
The World Health Organization, Avenue Appia 20, 1211, Geneva 27. Tel: (022) 91 21 11.

UK
Cromme, Derek, Dr, University of Bath, Avon.

Jones, Phillip, Dr, Welsh School of Architecture, PO Box 25, Cardiff, South Wales.

Olivier, David, Energy Advisory Associates, 156 Bradwell Road, Bradville, Milton Keynes, Buckinghamshire MK13 7AX.

Warren, Peter, Dr, Building Research Establishment, Backwells Lane, Garston, Watford, Hertfordshire WD2 7JR.

USA
US Environmental Protection Agency, 401 M St SW, Washington DC 20460.

Leaderer, Brian, Professor, Yale University, Pierce Laboratory, 290 Congress Avenue, New Haven, CT 95060.

Levin, Hal, Research Architect, 2548 Empire Grade, Santa Cruz, CA 95060. (Hal Levin is also Editor of *Indoor Air Quality Update*, published by Cutter Information Corp., 1100 Massachusetts Avenue, Arlington, MA 02174.)

National Bureau of Standards, Washington DC.

USSR
Mikulin, Alexander, 141093 Moscow Region, Bolshevo, Textilshik, Sovietskaya Street 22—86.

Jablokov, A.V., Moscow, pr. Kalinina 27, Ecological Commission.

The Environment by the numbers

In the USA:

1 20% of ground-water used for drinking is contaminated.
2 65,300 discharge pipes are legally dumping toxics in rivers etc.
3 94% of customers are not notified when drinking unhealthy water.
4 2,600,000,000 pounds of pesticides are used annually (10 lb/person).
5 313,000 US farm workers are affected by pesticide poison yearly.
6 85% of waste steam could be recycled.
7 200 air toxins are released annually.
8 The number of toxins regulated by the Environmental Protection Agency is seven.
9 3 billion gallons of petrol were wasted in 1984 due to traffic congestion.
10 One can of oil will contaminate 65,000 gallons of water.
11 257 million gallons of motor oil are poured into US water supplies yearly.
12 570,000 barrels of oil per day would be saved if all Americans lowered their thermostats by 6°F.

Source: *The Progressive Review*, No. 289, Washington DC, June 1990.

Endpiece:
Regaining a global
perception

There have been reports from Finland and the Soviet Union that people eating a lot of fish from waters with heavy algae blooms have died of toxic poisoning. Their deaths were probably due to their having eaten the fish roe and livers. This ties in with the views expressed in the illustration of using waste water to irrigate arable land in the section on external factors of influence (Section 4.1). Here the concern was with heavy metals leaching into ponds, and the knowledge that any toxic elements tend to collect in the liver and kidneys of fish, animals and people.

For the past three years blue-green algae blooms have become a particular hazard in Britain due to longer spells of stable weather. Although such toxic algae had been known for more than a century their effect in contaminating tap-water can lead to severe inflammation of the liver. One of the many toxins produced by the 25 different sorts of alga so far identified as poisonous is twice as deadly as cobra venom. The toxins are difficult to detect in the body, and many symptoms are put down to other causes. People affected in Britain have reported suffering from abdominal pains, vomiting, diarrhoea, blistering of the mouth and sore throats.

Using algicides to attack the blooms causes the release of even harsher toxins and allows the algae to return in a short time with renewed vigour.

There is no doubt that there are many agents of contamination that are either unknown or not yet classified as being dangerous. When people talk of the 'defence of the realm' they are usually thinking in military terms. But without proper resources both in government financing and specialist training we will find it difficult to protect outselves from sources that can be as dangerous to mankind as warfare. At no time previously in my life can I remember national and international concern about such things as mad cow disease and salmonella poisoning, on a scale that raises doubts about the way in which we have allowed ourselves to lose control over the means of producing food local to our needs in a sustainable and healthy way. The same problems and concerns are evident in the way in which we use our land for industry, in transport and in the way in which we build our cities and towns.

Our perception of the way in which we can create a healthy environment and design healthy buildings requires that we must be prepared to reach outside the talk and teaching of *distinct* parameters. Instead we must begin to look away from the frame of narrow definitions, and from being made to feel that one is a generalist by teachers and professional institutions who subscribe to the specialist view of knowing. Of course, in any profession one endeavours to become highly knowledgeable in several interrelated subjects. Few people are aware of the extent of the knowledge that building services engineers have to acquire and understand. To be able to give a client a full and comprehensive service of design advice one finds that the field of understanding becomes wider. It is with this widening of vision to understand the interactions of many other specialists that this book and the use of such matrix systems as ECHOES finds its place. It asks that engineers, architects and all other people concerned with buildings learn to become more holistic.

The use of any matrix system does, both in its formation as a grid and in its use, create constraints. Better would be the form of a circle or even an unended shape. But this does counter much of our Western thinking and some designers would think that I am dancing again. Throughout this book we have been looking at a new way of problem solving. But are WE the problem? Susan Coldicutt of the Department of Architecture at the University of Adelaide, Australia has been questioning the way we perceive problems.

She asks, 'To understand the way we frame or define problems it is essential to recognize that all applied knowledge is value-laden. While our task in making decisions about environments is to relate to the physical world of objects and environment to human and other ends, we can only do this by using knowledge which *represents* the physical world (or means) and knowledge which *represents* the ends which we seek to address. Inevitably, this knowledge of ends and means is not the same as the ends or means themselves'.

The same thinking can be applied to the way we think about energy. 'It is patently obvious that there is far less variation in the climatic conditions in any given building type (for instance a large commercial building) than there is in the climatic condition which the individual person is likely to experience throughout a 24-hour day. If climate control is to serve people's needs, how is it that people tolerate daily ranges from comfort to snow, or comfort to tropical outdoors? And if they tolerate this, on what grounds do decision-makers decide on extent of climate in various parts of the built environment?'[1]

It is clear that most theoretical explanations of people's thermal requirements ignore cultural, social, institutional and economic influences, or, at most, mention them in passing.

David Hockney, the painter, once said that if you look at a perspective painting by Canaletto, you, the individual, were excluded from the wholeness of things. You stood from without looking into a frame. You could see only what was in front of you. You could not look over the wall of a building and see what was happening beyond the confines of the frame.

Such western influences changed the way in which the Chinese saw things for thousands of years in their painting on a rolled scroll. Before the 1680s the Chinese allowed you to be at the centre of things. You were able to look all about you: to see what was over the walls; to see inside the people's houses; to witness many different events of daily life. You were not confined to one item of activity, tightly framed. There was a wholeness. Hockney believes that the loss of this perception has led to the gradual decline of the philosophical place that the Chinese had developed, and which we are now finding to have a bearing on our own ability to find a way through the manifold factors of influence that we need to deal with in the creation of a healthy building.

Forty years ago in the Egyptian desert the need to wash in clean water led to the making of a solar water heater from scrap. The need to find a cool internal environment led to the use in a simple building of a simple way to keep our drinking water cool and to provide a comfortable internal climate without an electric air-conditioning unit. Some 4000 years before Christ the people who lived on the River Euphrates built, on what we now call the 21° line, the city of Uruk. They completely recycled the water from the river in a way that cooled their buildings, as proposed in the Johannesburg scheme, and to a lesser degree in the NMB Bank, Amsterdam, as described in Chapter 5. These same people, who would wonder at some of our scientific creations, probably used reeds to clean the earth, and used waste water to irrigate their arable land, as talked about in Section 4.1. These people were at the centre of things: it is time that we are again.

May my house be in harmony
From my head, may it be happy
To my feet, may it be happy
Where I lie, may it be happy
All above me, may it be happy
All around me, may it be happy.
                                    Navajo*

*Source *The Night Chant* — *a Navajo ceremony*. Memoirs of the American Museum of Natural History Vol. VI, New York 1902.

Index